THE CASE AGAINST VACCINE MANDATES

Other Books by Kent Heckenlively

Ending Plague (with Dr. Francis W. Ruscetti and Dr. Judy A. Mikovits)
Behind the Mask of Facebook (with Ryan Hartwig)
Google Leaks (with Zach Vorhies)
Inoculated
The Case for Interferon (with Joseph Cummins)
The Case Against Masks (with Dr. Judy A. Mikovits)
Plague of Corruption (with Dr. Judy A. Mikovits)
Plague (with Dr. Judy A. Mikovits)

THE CASE AGAINST VACCINE MANDATES

KENT HECKENLIVELY, JD

HOT BOOKS

Hot Books may be purchased in bulk at special discounts for sales promotion, corporate gifts, fund-raising, or educational purposes. Special editions can also be created to specifications. For details, contact the Special Sales Department, Skyhorse Publishing, 307 West 36th Street, 11th Floor, New York, NY 10018 or info@skyhorsepublishing.com.

Hot Books® and Skyhorse Publishing® are registered trademarks of Skyhorse Publishing, Inc., a Delaware corporation.

Visit our website at skyhorsepublishing.com.

10 9 8 7 6 5 4 3 2

Library of Congress Cataloging-in-Publication Data is available on file.

Print ISBN: 978-1-5107-7103-1
Ebook ISBN: 978-1-5107-7105-5

Printed in the United States of America

Dedicated to Saint Michael: The greatest warrior
in God's heavenly army and humanity's best friend.

[T]he people can always be brought to the bidding of the government. All you have to do is tell them they are being attacked and denounce the pacifists for lack of patriotism and exposing the country to danger. It works the same in any country.
—Nazi leader Hermann Goering, in an interview with Gustave Gilbert from his jail cell during the Nuremberg trials, April 18, 1946

If you're not careful, the newspapers will have you hating the people who are being oppressed and loving the people who are doing the oppressing.
—Malcom X

Contents

Defy the Government and Get a Five Dollar Fine—*Jacobson v. Massachusetts* (1905) and How Vaccine Mandates Got Started

It seems to me the decision by any governmental body as to whether it should pursue a vaccine mandate is properly premised upon the answers to a few simple, but important questions.

The first is how much power a state or nation actually has over a person's bodily integrity, based upon previously existing law.

The second is, when such power has been asserted by the state, what have been the results?

The final question is, despite any prejudices people might have on these questions, can they be expected to possess the intellectual integrity necessary to change their minds when presented with new evidence or facts that they had not previously known?

I hope I have your agreement that this is a reasonable framework for us to begin our discussion.

* * *

Most legal experts, including Harvard law professor Alan Dershowitz, cite the 1905 United States Supreme Court decision in *Jacobson v. Massachusetts* for the proposition that governments can coerce vaccine compliance, as he did in an exchange with Fox TV host Laura Ingraham on July 29, 2021. From an article by *Yahoo News* on the exchange:

> Fox TV host Laura Ingraham locked horns with lawyer Alan Dershowitz over the prospect of mandated vaccination.
>
> Mr. Dershowitz was arguing the case for compulsory vaccination and made a comparison between COVID-19 and smallpox on *The Ingraham Angle*.
>
> He said: "As far as mandating vaccination, I think the Supreme Court would uphold gradual mandating of vaccination. First, conditioning going to school on getting vaccinated, conditioning getting on airplanes, conditioning getting in crowded buildings."
>
> The former Harvard Law School professor continued, telling Ms. Ingraham that George Washington had mandated vaccination against smallpox for his troops during the Revolutionary War.[1]

As far as setting up the terms of the debate, the article did a fairly good job. Professor Dershowitz was on the side of the government mandating a COVID-19 vaccine and Fox News host Laura Ingraham was against it.

In the next few paragraphs, the article describes Ingraham as an individual who has taken "a consistent anti-vax stand throughout the pandemic" and disputed the association of COVID-19 with historical smallpox outbreaks. Ingraham noted the COVID-19 vaccines were not even approved, which drew a sharp rebuke from professor Dershowitz:

> "Neither was the smallpox vaccine in 1905," retorted Mr. Dershowitz, referencing the United States Supreme Court case

Jacobson v. Massachusetts about smallpox vaccines, in which the court upheld the authority of states to enforce compulsory vaccination laws.

He went on: "I think Covid is worse than smallpox in many ways. It may not kill as many people but we don't know what the long-term impact is."

"It's killed 300 million people worldwide," said Ms. Ingraham, meaning smallpox.[2]

Dershowitz continued his argument, asserting his right to board an airplane without the fear that somebody might infect him with the virus. At this assertion, Ingraham started laughing and replied:

"Professor, have you not been listening?" she asked. "I may not have gone to Harvard Law School, but I did hear the president today talk about how if you're vaccinated you still can spread the virus. The data out of Israel, the data out of the UK, they're freaking out about this."

"It'll be spread much less seriously," replied Mr. Dershowitz.

"You can't deprive people of their constitutional rights on the basis of a vaccine that still allows the spread of the virus. Okay," said Ms. Ingraham, before wrapping up the segment.[3]

For the vast majority of the public, that's how the debate is framed, and how each side argues their position. Ingraham argues that COVID-19 is not smallpox, while Dershowitz replies that he believes COVID-19 will be worse than smallpox. Dershowitz asserts the right to board an airplane without the fear of contracting the virus. Ingraham notes the most recent data that the vaccine is NOT effective at stopping the spread of the virus. Dershowitz counters by claiming that the virus spread by vaccinated individuals will be much less severe than the virus spread by unvaccinated individuals.

I think any fair-minded person would agree that when the debate is framed in this manner, there are strong arguments on both sides.

However, as I read the initial case of *Jacobson v. Massachusetts*, as well as commentary on the case, I couldn't help but believe that both Ingraham and Dershowitz have misinterpreted the original case which allegedly gave the government such draconian power.

Let us return to the America of 1905, a country which at that time denied women the right to vote and upheld racial and religious discrimination, and determine, if we can, what the Supreme Court believed to be the proper balance between the power of the state and the right of an individual to make their own health decisions.

* * *

If one were to simply read the opinion in *Jacobson v. Massachusetts*, it would seem to support the position of Mr. Dershowitz and even contain a few surprises to modern day readers. For example, in the syllabus of the opinion (a summary of the main points often provided at this time), it notes specifically that the preamble to the US Constitution has no place in American law:

> The United States does not derive any of its substantive powers from the Preamble of the Constitution. It cannot exert any power to secure the declared objects of the Constitution unless, apart from the preamble, such power can be found in, or properly implied from, some express delegation in the instrument.[4]

Pastor Henning Jacobson was relying on the preamble to the US Constitution to support his claim that the fine for being unvaccinated was a violation of his civil rights. For those unable to recite the preamble to the Constitution from memory, I will provide it for you:

> We the People of the United States, in Order to form a more perfect union, establish justice, insure domestic tranquility, provide

for the common defense, promote the general Welfare, and secure
the blessings of Liberty to ourselves and our Posterity, do ordain
and establish this Constitution for the United States of America.[5]

In effect, the pastor was declaring that he was a free person, able to
make his own decisions under the United States Constitution as well as
under the provisions of the Fourteenth Amendment, which provided:

No state shall make or enforce any law which shall abridge the
privileges or immunities of citizens of the United States; nor shall
any State deprive any person of life, liberty, or property without
due process of law; nor deny to any person within its jurisdiction
the equal protection of the laws.[6]

The court laid out the series of events which had led to the filing of a
complaint against Pastor Jacobson and its resolution.

The complaint charged that, on the seventeenth day of July, 1902,
the Board of Health of Cambridge, being of the opinion that it
was necessary for the public health and safety, required the vaccination and revaccination of all the inhabitants therof who had
not successfully vaccinated since the first day of March, 1897, and
provided them with the means of free vaccination, and that defendant, being over twenty-one years of age and not under guardian-
ship, refused and neglected to comply with such requirement.

The defendant, having been arraigned, pleaded not guilty.
The government put in evidence the above regulations adopted
by the Board of Health and made proof tending to show that its
chairman informed the defendant that, by refusing to be vaccinated, he would incur the penalty provided by statute, and would
be prosecuted therfore; that he offered to vaccinate the defendant
without expense to him, that the offer was declined, and defendant refused to be vaccinated.

The prosecution having introduced no other evidence, the defendant made numerous offers of proof. But the trial court ruled that each and all of the facts offered to be proved by the defendant were immaterial, and excluded all proof of them.

The defendant, standing upon his offers of proof and introducing no evidence, asked numerous instructions to the jury, among which were the following:

That section 137 of chapter 75 of the revised Laws of Massachusetts was in derogation of the rights secured to the defendant by the Preamble to the Constitution of the United States, and tended to subvert and defeat the Purposes of the Constitution as declared in its Preamble.

That the section referred to was in derogation of the rights secured to the defendant by the Fourteenth Amendment of the Constitution of the United States, and especially of the clauses of that amendment providing that no States shall make or enforce any law abridging the privileges and immunities of citizens of the United States, nor deprive any person of life, liberty or property without due process of law, nor deny to any person within its jurisdiction the equal protection of the laws; and that said section was opposed to the spirit of the Constitution.

Each of the defendant's prayers for instructions was rejected, and he duly excepted. The defendant requested the court, but the court refused, to instruct the jury to return a verdict of not guilty. And the court instructed the jury, in substance, that, if they believed the evidence introduced by the Commonwealth and were satisfied beyond a reasonable doubt that the defendant was guilty of the offense charged in the complaint, they would be warranted in finding a verdict of guilty. A verdict of guilty was thereupon returned.

The case was then continued for the opinion of the Supreme Judicial Court of Massachusetts. The court overruled all the defendant's exceptions, sustained the action of the trial court, and thereafter, pursuant to the verdict of the jury, *he was sentenced by the court to pay a fine of five dollars. And the court ordered that he stand committed until the fine was paid.* [Bold and italics added by author.][7]

There are many things we can take away from this account, not the least of which was that Pastor Jacobson was a very persistent individual.

But probably the most shocking finding was the penalty which was assessed for his refusal to be vaccinated. He was fined a grand total of five dollars, which in today's money is about $150. (Less than one third the cost of blowing through a red light in California, where I live.)

Could it really be that after all the sound and fury of the vaccine debate when smallpox was ravaging the world that the only penalty which would be assessed against a person who declined a smallpox vaccine would be a monetary penalty less significant than many modern-day traffic tickets?

In 2005, an inter-disciplinary group of scientists and lawyers from Boston University published an article in the *American Journal of Public Health* assessing the importance of the Jacobson case and what it might mean in the twenty-first century. The authors wrote:

> As the 20th century began, epidemics of infectious diseases such as smallpox remained a recurrent threat. A Massachusetts statute granted city boards of health the authority to require vaccination "when necessary for public health or safety." In 1902, when smallpox surged in Cambridge, the city's board of health issued an order pursuant to this authority that required all adults to be vaccinated to halt the disease. The statutory penalty for refusing vaccination was a monetary fine of $5 (about $100 today). *There was no provision for actually forcing vaccination on any person.* [Bold and italics added by author.]
>
> Henning Jacobson refused vaccination, claiming that he and his son had bad reactions to earlier vaccinations. The

Massachusetts Supreme Judicial Court found it unnecessary to worry about any possible harm from vaccination, because no one could actually be forced to be vaccinated: "If a person should deem it important that vaccination should not be performed in his case, and the authorities should think otherwise, it is not in their power to vaccinate him by force, and the worst that could happen to him under the statute would be the payment of $5."[8]

Considering the vehemence of today's arguments over vaccine mandates, and whether individuals should be barred from social gatherings such as restaurants, bars, or movies, or even fired from their jobs as a result of declining a COVID-19 vaccination, the Jacobson decision seems like an echo of a long-vanished era which valued individual rights in a manner which seems to be quickly vanishing in our modern world.

In her debate with Alan Dershowitz, Laura Ingraham was correct in her assertion that smallpox has been responsible for more than 300 million deaths.[9] But how different might their discussion have been if both of them were aware that even in the face of such unprecedented dangers, the state government of Massachusetts and the United States Supreme Court were unwilling to violate Henning Jacobson's bodily integrity or conscience rights?

In many realms, such as women's health and patient privacy laws, there has been an increased attention to personal freedoms. In 1905, the prospect of women having the right to vote was a distant hope, and the final demolition of Jim Crow segregation laws were more than half a century away. Yet the individual was in charge of his or her own body, without the threat of banishment from society, the right to travel, or being fired from their job because of their personal medical decisions. Have we gone backwards in our protection of liberty and conscience?

I believe, as the interdisciplinary team of scientists and lawyers from Boston University do, that:

> One practical reason for protecting constitutional rights is that it encourages social solidarity. People are more likely to trust officials who protect their personal liberty. Without trust, public officials will not be able to persuade the public to take even the most reasonable precautions during an emergency, which will make a bad situation even worse.
>
> The public will support reasonable public health interventions if they trust public health officials to make sensible recommendations that are based on science and where the public is treated as part of the solution instead of the problem. Public health programs that are based on force are a relic of the 19th century; 21st-century public health depends on good science, good communication, and trust in public health officials to tell the truth. In each of these spheres, constitutional rights are the ally rather than the enemy of public health. Preserving the public's health in the 21st century requires preserving respect for personal liberty.[10]

I hope I've given enough information to convince you that the *Jacobson v. Massachusetts* case presented to the public by both the left and the right bears little resemblance to the actual facts of the case.

While I agree that if one were to merely skim the case for a couple choice quotes, one might be forgiven for claiming that the decision gave near God-like powers of choice to the government for healthcare decisions. But it's a much more nuanced decision.

To be fair, the government does claim for itself the right to make healthcare decisions during a time of crisis, and can even enforce such suggestions by a fine which is in the range of many modern-day traffic tickets. As a matter of principle, I would continue to argue that this approach is wrong from a constitutional perspective, in which all powers not expressly ceded to the government are maintained by the individual.

However, as a practical matter, I doubt anybody would have much trouble living with a government which operated under Jacobson principles. The cost of following one's conscience would be relatively low and society would be more peaceful if all lived according to the dictates of their conscience.

I just wish that Laura Ingraham and Alan Dershowitz had looked more closely at the facts of the Jacobson case.

CHAPTER TWO

Buck v. Bell (1927)—Why We Sterilized Imbeciles and the Rise of the American Eugenics Movement

Has there ever been a case in American history which was so thoroughly misunderstood as *Jacobson v. Massachusetts* and has caused so much damage, not just in this country, but around the world as well?

Jacobson believed it was his right to avoid a vaccination which he believed would cause him harm. The state of Massachusetts believed differently. However, the remedy was not to grab the good pastor and stab a needle in his arm, but to give him a five dollar fine.

As I've stated before, I do not believe the state has the authority to compel any medical treatment, just as they do not have the right to force a vegan to have a hamburger because the state is concerned they are not getting enough vitamin B-12. It is the God-given right of each human being to make their own decisions about their health.

With that being stated, I believe the compromise of Jacobson is something that most people could live with, as it allowed Pastor Jacobson to follow his conscience, and yet also acknowledged the power of the state.

However, the danger with such compromises is that they can then be used as a wedge in the next case to bring about unimaginable suffering. And with that in mind we come to the 1927 US Supreme Court case of *Buck v. Bell, Superintendent of State Colony Epileptics and Feeble Minded*:

> Carrie Buck is a feeble-minded white woman who was committed to the State Colony above mentioned in due form. She is the daughter of a feeble-minded mother in the same institution, and the mother of a feeble-minded child. She was eighteen years old at the time of the trial of her case in the Circuit Court, in the latter part of 1924. An Act of Virginia approved March 20, 1924 (Laws 1924, c. 394) recites that the health of the patient and the welfare of society may be promoted in certain cases by the sterilization of mental defectives, under careful safeguard, etc.; that the sterilization may be effected in males by vasectomy and in females by salpingectomy, without serious pain or substantial danger to life; that the commonwealth is supporting in various institutions many defective persons who if now discharged would become a menace, but if incapable of procreating, might be discharged with safety and become self-supporting with benefit to themselves and to society; and that experience has shown that heredity plays an important part in the transmission of insanity, imbecility, etc.
>
> The statute thus enacts that whenever the superintendent of certain institutions, including the abovenamed State Colony, shall be of opinion that it is for the best interest of the patient and society that an inmate under his care should be sexually sterilized, he may have the operation performed upon any patient afflicted with hereditary forms of insanity, imbecility, etc, on complying with the very careful provisions by which the act protects the patients from possible abuse.[1]

Simply reading the language of this 1927 United States Court opinion in 2021 is likely to produce revulsion in the mind of the average reader. People using expressions such as "feeble-minded" or "mental defectives" or "imbecility" would immediately find themselves dealing with a torrent of condemnation. When this was the language which described psychiatric patients by the United States Supreme Court, one shudders to imagine how these poor souls were likely treated in these institutions by their caregivers.

By the laws passed by the State of Virginia, and upheld by the US Supreme Court, the superintendent of these institutions could decide to sterilize his male and female patients. You may ask what connection the involuntary sterilization has to do with the *Jacobson* case and vaccines, but *Jacobson* was the *sole* case cited to support the involuntary sterilization of Carrie Buck. In probably one of the most horrific passages of any Supreme Court decision, Justice Holmes wrote:

> We have seen more than once that the public welfare may call upon the best citizens for their lives. It would be strange if it could not call upon those who already sap the strength of the State for these lesser sacrifices, often not felt to be such by those concerned, in order to prevent our being swamped with incompetence. It is better for all the world, if instead of waiting to execute dangerous offspring for crime, or to let them starve for imbecility, society can prevent those who are manifestly unfit from continuing their kind. ***The principle that sustains compulsory vaccination is broad enough to cover cutting the Fallopian tubes.*** Jacobson v. Massachusetts, 197 U. S. 11, 25 S. Ct. 358, 49 L. Ed. 643, 3 Ann. Cas. 765. ***Three generations of imbeciles are enough.*** [Italics and bold added by author.][2]

Justice Oliver Wendell Holmes Jr. is considered one of the greatest minds to ever serve on the United States Supreme Court and is generally known for his judicial restraint. While this philosophy is generally applauded as allowing the legislature and, by extension, the citizens, to determine the laws, there are times when this approach can go wildly off kilter, especially when powerful interests can more effectively move the levers of government than the people.

However, sometimes this approach carries with it the danger that some truly tyrannical actions may be allowed to occur. Let's see if we can follow his argument, and more importantly, whether we agree with it.

Holmes quite correctly notes that the government can call upon its citizens and ask for their lives, as is often done during wartime. Holmes himself was an army veteran, having fought on the Union side during the American Civil War. The idea that a government may call upon its citizens to fight in a war would find wide agreement.

But it does not seem to follow that a government may "call upon those who already sap the strength of the State for these lesser sacrifices," which in this example consists of involuntary sterilization. Holmes assumes that since he is confident that the offspring of such individuals will commit crimes or be a burden upon the State, he may simply prevent them from ever being born.

This sounds to me like the justification used by every tyrant in history for their reign of brutality and terror. The tyrant has greater vision than those of his subjects and is thus freed from the bounds of conventional morality in order to commit unspeakable crimes.

Buck v. Bell gave free reign to involuntary sterilization practices and, by 1932, twenty-nine states had passed laws allowing for the practice.[3] These laws eventually resulted in the involuntary sterilization of more than sixty thousand Americans before these practices were slowly abandoned after World War II.[4] Probably the most shocking thing to realize is that most of these procedures took place in what we consider progressive states, such as California, which accounted for roughly one-third of all sterilizations and did not remove its involuntary sterilization laws from the books until 1979.[5]

These laws were passed and these procedures were not undertaken by accident. They were the result of a philosophical movement which reached into the highest levels of medicine and believed in the ability of science to reengineer human beings into becoming more compliant citizens of the state. I am not speaking of a movement in Nazi Germany, or Soviet Russia, but within the borders of the United States.

I am speaking of the American eugenics movement.

Let me tell you some of their shameful history.

* * *

As one writer on the eugenics movement recounted:

> The eugenics movement will forever be associated with Adolph Hitler, whose quest to build an Aryan master race during the 1930s and '40s culminated in the extermination of millions.
>
> However, Hitler wasn't the first to champion the idea of wiping away humans deemed to be unfit. In large part, he actually took inspiration from the United States. As Hitler remarked in 1924's *Mein Kampf*, "There is today one state in which at least weak beginnings toward a better conception are noticeable. Of course, it is not our model German Republic, but the United States."
>
> The popularity of eugenics and related ideas in the U.S. (as well as Western Europe) at the time was in part a reactionary response to increased industrialization and immigration. The latter was on the rise as cities became more crowded as people moved to be closer to work. And with supporters of the early eugenic movement believing that people inherited traits like feeble-mindedness and poverty, this meant that society had an obligation to thin this growing herd.[6]

Now we have some context to understand the opinion of US Supreme Court Justice Oliver Wendell Holmes Jr.

Holmes was a supporter of eugenics, as was a great portion of the intellectual elite of our country. This will become abundantly clear as we continue. A little more context for the genesis and popularity of eugenics among the progressive class is probably necessary at this point:

> Fittingly enough, eugenics actually has some of its roots with Charles Darwin. His theories about "survival of the fittest" inspired his cousin, Francis Galton, to start the eugenics movement as the world would come to know it . . .
>
> . . . Both abroad and in the United States, proponents of the eugenics movement believed it a Caucasian responsibility to Westernize other civilizations. This was coupled with the idea of producing fewer, *better* children to create a better race, and cure many economic and social problems . . .
>
> . . . Presidents, economists, activists, and philosophers—many of which you'd never think would be supporters—all once spoke out in support of eugenics.[7]

It should now be clear that eugenics was a philosophy developed by those who would have described themselves as "progressive intellectuals," comfortable with making decisions on behalf of other individuals.

While it is no doubt true that eugenics was embraced by a broad spectrum of our country's intellectual elite, there can be little debate that such beliefs were in stark opposition to the founding ideals of Western civilization, namely that each person possesses God-given rights to conscience and the ability to live their life as they determine.

Although the poisonous idea of eugenics was born in England and took root among substantial parts of the American public, it found its most fertile ground in Hitler's Germany. The clash between the ideals of Western civilization and Hitler's Germany were thus inevitable, and in the peace which followed, much of the intellectual elite's past support of eugenics was wiped from the public consciousness. However, in 2016, *Harvard Magazine* published a long article on the university's past support of eugenics, specifically naming Oliver Wendell Holmes

Jr. as one of the leading figures in the movement, and it makes for frightening reading:

> In August 1912, Harvard president emeritus Charles William Eliot addressed the Harvard Club of San Francisco on a subject close to his heart: racial purity. It was being threatened, he declared, by immigration. Eliot was not opposed to admitting new Americans, but he saw the mixture of racial groups it could bring about as a grave danger. "Each nation should keep its stock pure," Eliot told his San Francisco audience. "There should be no blending of races."
>
> Eliot's warning against mixing races—which for him included Irish Catholics from marrying Anglo-Saxon Protestants, Jews marrying Gentiles, and blacks marrying whites—was a central tenet of eugenics.[8]

To put it plainly, the former president of Harvard University did not believe "diversity is our strength" and had little hesitation in telling that to his San Francisco audience. One wonders if this shocking racism was as widespread among the working class, who were more likely to live in close proximity to and work with individuals of different races and cultures. Much has been made of the disdain many in the working class have for the intellectual elite, with strong claims being made of "anti-intellectualism." And yet, when one considers that the highly educated of earlier times were quite comfortable with the idea of managing, and even culling, the members of the lower classes, whom they deemed unacceptable, such disdain is not only warranted, but a survival instinct.

The article went on:

> The former Harvard president was an outspoken supporter of another major eugenics cause of his time: forced sterilization of people declared to be "feeble-minded," "physically disabled,"

"criminalistic," or otherwise flawed. In 1907, Indiana had enacted the nation's first eugenic sterilization law. Four years later, in a paper on "The Suppression of Moral Defectives," Eliot declared that Indiana's law "blazed the trail which all free states must follow, if they would protect themselves from moral degeneracy."

He also lent his considerable prestige to the campaign to build a global eugenics movement. He was a vice president of the First International Eugenics Congress, which met in London in 1912 to hear papers on "racial suicide" among Northern Europeans and similar topics. Two years later, Eliot helped organize the First National Conference on Race Betterment in Battle Creek, Michigan.[9]

Perhaps my strong reaction to the comments of the former president of Harvard University comes from my "moral degeneracy" as a result of the "racial suicide" in which my ancestors clearly engaged in for hundreds of years. My 23andMe results show my two strongest ancestry contributions are from Sicily (51.3 percent), the southernmost part of Italy, and Norbotten County, Sweden (24.6 percent), the most northern county of Sweden, with part of it lying just inside the Arctic Circle. Already I've messed up my racial purity by having ancestors from the far north and the far south of Europe. But it gets even worse. The rest of my European bloodline is British and Irish, then French and German, and finally, to finish it all off, I'm 3.4 percent Finnish.

But don't think, well, yeah, he's just European. Big deal.

My 23andMe results show that between 1690 and 1840 I had an ancestor who was Iranian, and one who was Middle Eastern. In a smaller frame of time, between 1690 and 1810, I had another ancestor who was 100 percent North African.

I proudly declare myself as a child from the four corners of Europe, as well as Africa and Asia. In total, my genetic ancestry shows I have the blood of three different continents flowing in my veins.

It seems President Eliot was doing just about everything within his power to prevent a person like me from ever coming into the world. And that Harvard was the center of this perverted universe will probably come as a surprise to some:

> Harvard's role in the movement was in many ways, not surprising. Eugenics attracted considerable support from progressives, reformers, and educated elites as a way of using science to make a better world. Harvard was hardly the only university that was home to prominent eugenicists. Stanford's first president, David Starr Jordan, and Yale's most acclaimed economist, Irving Fischer, were leaders in the movement. The University of Virginia was a center of scientific racism, with professors like Robert Bennet Bean, author of such works of pseudo-science as the 1906 *American Journal of Anatomy* article, "Some Racial Peculiarities of the Negro Brain."
>
> But in part because of its overall prominence and influence on society, and in part because of its sheer enthusiasm, Harvard was more central to American eugenics than any other university.[10]

President Eliot was not alone in his eugenics campaign. Oliver Wendell Holmes Sr., the father of Supreme Court Justice Oliver Wendell Holmes Jr., was once the dean of Harvard Medical School and was "one of the first American intellectuals to espouse eugenics."[11] It seems that hatred for the "defectives" (perhaps known today as the "deplorables") runs in the family.

Holmes Sr. had made his first public declaration of support for eugenics in 1875,[12] and it was in 1927 that Holmes Jr. authored the majority opinion in *Buck v. Bell*. It was only in 1979 that California, one of the last holdouts, finally repealed its involuntary sterilization laws.

Bad ideas, especially when supported by progressive forces in academia and government, can often take a long time to die.

* * *

It wasn't just at Harvard or on the US Supreme Court that eugenic theory thrived and was put into practice. Eugenics made its way into scientific circles and was supported by some of the wealthiest men of the time:

> Eugenics would have been so much bizarre parlor talk had it not been for extensive financing by corporate philanthropies, specifically the Carnegie Institution, the Rockefeller Foundation, and the Harriman railroad fortune. They were all in league with some of America's most respected scientists hailing from such prestigious universities as Stanford, Yale, Harvard, and Princeton. These academicians espoused race theory and race science, then faked and twisted data to serve eugenics' racist aims.
>
> Stanford president David Starr Jordan originated the notion of "race and blood" in his 1902 racial epistle "Blood of a Nation," in which the university scholar declared that human qualities and conditions such as talent and poverty were passed in the blood.[13]

When we think of philanthropic organizations of the past, we do not think of them supporting the eugenics program which was enthusiastically adopted by the Nazis for their planned murder of millions. However, when we view what these "philanthropists" were doing in the past, which has escaped the requisite public attention, it's perhaps understandable why people question the motivations of some of our modern-day philanthropists like Bill Gates or groups such as the World Health Organization.

The center of the American eugenics program for scientists was without a doubt the Cold Spring Harbor Lab in Long Island, New York. However, if you went to the front page of their website, you'd find the following short description of the lab's history:

> Founded in 1890, Cold Spring Harbor Laboratory has shaped contemporary biomedical research and education with programs

in cancer, neuroscience, plant biology, and quantitative biology. Home to eight Nobel Prize winners, the private, not-for-profit Laboratory employs 1,000 people, including 600 scientists, students, and technicians. The Meetings and Courses Program hosts more than 12,000 scientists from around the world each year on its campuses in Long Island and Suzhou, China. The Laboratory's education arm also includes an academic publishing house, a graduate school, and programs for middle and high school teachers and students.[14]

Perhaps it's too much for an institution to highlight the crimes of which it was guilty decades earlier. But it is a mystery how the "progressivism" of the early twentieth century found such common areas of agreement with the ideals of Hitler's Germany.

A *New York Times* article from 2014 provides a more accurate picture of the history of the Cold Spring Harbor Laboratory:

> An old stucco house stands atop a grassy hill overlooking the Long Island Sound. Less than a mile down the road, the renowned Cold Spring Harbor Laboratory bustles with more than 600 researchers and technicians . . .
>
> But that old house, now a private residence on the outskirts of town, once held a facility whose very name evokes dark memories: the Eugenics Record Office.
>
> In its heyday, the office was the premier scientific enterprise at Cold Spring Harbor. There, bigoted scientists applied rudimentary genetics to singling out supposedly superior races and degrading minorities. By the mid-1920s, the office had become the center of the eugenics movement in America.[15]

It's a common tactic to try to rewrite history, but to its credit, the *New York Times* article lays out the facts. At one time, the eugenics program was "the premier scientific enterprise at Cold Spring Harbor."

It's difficult to get more definitive than that. In addition, by the mid-1920s, Cold Spring Harbor had become the "center of the eugenics movement in America."

Think about what that means.

The facility today houses eight Nobel Prize winners. It's difficult to imagine any claim that the eugenics program and its support by some of the most powerful academic and financial forces in the country was some well-intentioned mistake.

The *New York Times* article continued:

> When the Eugenics Record Office opened its doors in 1910, the founding scientists were considered progressives, intent on applying classic genetics to breeding better citizens. Funding poured in from the Rockefeller family and the Carnegie Institution. Charles Davenport, a prolific Harvard biologist, and his colleague, Harry H. Laughlin, led the charge . . .
>
> By the 1920s, the office had begun to influence the United States government. Laughlin testified before Congress, advocating forced sterilization and anti-immigration laws. The Immigration Act of 1924 effectively barred Eastern Europeans, Jews, Arabs and East Asians from entering the country. And, at the state level, thousands of people who were deemed unfit were sterilized.
>
> The University of Heidelberg in Nazi Germany later awarded Laughlin an honorary degree for his work in the science of "racial cleansing." He accepted the award, and his research on Long Island continued to influence Nazi ideology throughout World War II and the Holocaust.[16]

It's shocking to realize the ideas of racial superiority utilized in Hitler's Germany were to a great extent developed by two Harvard University professors. When large swaths of Americans today believe that many in the political, financial, and academic elite hate the common person and would be more than happy to see them eliminated, it's difficult not to look at this chapter in history and see a confirmation of many of their worst fears.

When the Nazi government essentially says, "Hey, good job, Harvard, with your racial superiority research, here's an honorary degree from one of our top universities!" you know you've passed into some very dark territory.

I think it's very important to look at the collection of forces which eugenics had arrayed by the 1920s, because in it we might see something similar to the collection of forces that are currently arrayed in supporting the idea of mandatory vaccination.

* * *

Starting in 1875, you've got Oliver Wendell Holmes Sr., the dean of Harvard Medical School, proclaiming his support for the eugenics theory of Darwin's cousin, Francis Galton. Little Oliver Wendell Holmes Jr., the future US Supreme Court justice who in 1927 will write the majority opinion in *Buck v. Bell*, grows in the searing cauldron of this racial hatred. As the years pass, others answer the siren call of eugenics, such as former Harvard president Charles William Eliot in 1912, and the vast fortunes of the Rockefeller and Carnegie family which lend their financial support. In 1910, the eugenics movement has one of its greatest successes, establishing a laboratory at Cold Spring Harbor, establishing a veneer of scientific credibility.

And with this power they start to lobby Congress, no doubt accompanied by a media influence campaign.[17] Although this effort meets my own personal definition of evil, one has to admire its brilliance the way one might marvel at the plan of some comic book villain in a superhero movie.

These people are not stupid.

The Congressional influence campaign bore fruit, as they passed the 1924 Anti-Immigration Act. I understand we have our own arguments about immigration today. But I must point out that the immigration arguments posed by the eugenicists of the 1920s were of a fundamentally different character.

Today's argument is over how much we should battle "illegal immigration" and whether our southern border is "secure."

Nobody is arguing today that certain races should not be allowed entry into our country. They would be shouted down and driven from the public square.

In addition to preventing certain ethnic groups from entering our country, the eugenics program sought to specifically harass those same ethnic groups within our country:

> The Harriman railroad fortune paid local charities, such as the New York Bureau of Industries and Immigration, to seek out Jewish, Italian, and other immigrants in New York and other crowded cities and subject them to deportation, trumped up confinement or forced sterilization.
>
> The Rockefeller Foundation helped found the German eugenics program and even funded the program that Josef Mengele worked in before he went to Auschwitz.
>
> Much of the spiritual guidance and political agitation for the American eugenics movement came from California's quasi-autonomous eugenic societies, such as the Pasadena-based Human Betterment Foundation and the California branch of the American Eugenics Society, which coordinated much of their activity with the Eugenics Research Society in Long Island. [Cold Spring Harbor].[18]

The American eugenics program then is best understood as a criminal enterprise akin in many ways to organized crime with several moving parts. There is the major financial and academic support, which seems to be centered mainly in the wealthy, progressive states such as Massachusetts (home of Harvard University), New York (Harriman and Rockefeller fortunes, as well as Cold Spring Harbor Laboratory), and California (Stanford University and many of the most prominent eugenics societies). By establishing their power bases, this movement then sought to influence the courts and our government, which they did quite successfully through a campaign which sought to stoke fear about the harm which would be done to our country if we allowed too many Italians, eastern Europeans, Jews, Arabs, or Asians into our country.

When the groups opposing mandatory vaccination point out the similarities between the eugenics campaign of the 1920s and '30s and the effort to mandate vaccines, it's because the forces pushing them are the same, with the possible exception of the financial strength that the Big Pharma companies can add to the effort. In effect, Big Pharma, along with billionaire Bill Gates, has taken over the role once played by the Rockefeller, Harriman, and Carnegie foundations.

In fact, let's even go to Bill Gates' own *Microsoft News* for their estimation of how much money Big Pharma will make from COVID-19 vaccines. I mean, if you can't trust the news network founded by Bill Gates, who can you trust for independent, objective reporting?

> Manufacturing and distributing effective vaccinations against COVID-19 has been heralded as the only way out of the pandemic and the race is on to get as many doses produced as possible. With demand through the roof, producing these life-saving drugs has become a lucrative business and the global coronavirus vaccine market will reach an enormous value of $35 billion in 2021 and is expected to grow to $42 billion by 2022.[19]

There you have it in black and white from *Microsoft News*. In 2021 the global market for COVID-19 vaccines will reach $35 billion dollars and, in 2022, it will reach $42 billion dollars.

With so much money on the line, do you really think you're getting the truth about the safety of these vaccines?

* * *

Who are the anti-vaxxers?

Well, it's a good thing to know that the Centers for Disease Control and Prevention has researched this question and come up with answers. As reported in the *Philadelphia Inquirer* in April 2019:

> If you're against vaccinating your child, there's a good chance you're a college-educated white woman making decent money.
>
> The rebel forces in America's latest culture war—the so-called anti-vaxxers—are often described as middle-and-upper class women who breast-feed their children, shop at Whole Foods, endlessly scour the web for vaccine-related conversation, and believe that their thinking supersedes that of doctors. Typically, their families make more than $75,000 dollars a year.
>
> That's based on findings from various studies, including the National Immunization Surveys conducted by the federal Centers for Disease Control and Prevention's National Center for Immunization and Respiratory Diseases.[20]

As an attorney, you learn that framing an issue, and the narrative you offer of a situation, is of critical importance. Because I'm trained in that branch of knowledge, it allows me to understand the framing and narrative provided by others. Let me give my reaction to the framing and narrative of the *Inquirer* article. The author wants you to be dismissive of these college-educated, white, working women, who are either middle or upper class.

In essence, the writer is saying to you, "Yeah, sure these women have a college degree, maybe even more, and are succeeding professionally. But we can't really call them smart."

However, when I read this, I have a different framing and narrative. I say to myself, middle to upper class, college educated: it probably means they're the first or second generation to go to college, but they've still got the common sense of the working class. Three generations ago, they would've been the people the eugenicists of Harvard and Stanford wanted to keep out of the country.

In my framing, I'd probably say these were the first people who picked up on the lies told by our government about the Vietnam War and protested. The first to realize President Nixon was a crook. The first to realize President Reagan was selling arms to the Iranians and diverting the proceeds to the Contra rebels in Nicaragua. The first to realize President Clinton was having sex with "that woman," his intern, Monica Lewinsky. The first to question whether the surveillance provisions of the Patriot Act would eventually be turned on American citizens. The first to question the intelligence reports that Iraq had weapons of mass destruction. And today, in addition to questioning the safety of the COVID-19 vaccines, the first to be skeptical of the integrity of the 2020 presidential election.

That's the way I'd frame it. You may think differently and that's fine. The *Inquirer* article continued:

> Doctors don't mince words about anti-vaxxers.
>
> "Frankly, these Caucasian, suburban, educated parents believe they can Google the word vaccine and get as much information as anybody," said Paul Offit, a professor of pediatrics and director of the vaccine education center at Children's Hospital of Philadelphia.
>
> "These people are educated just well enough to make terrible decisions for their children."[21]

When I read the comments of Dr. Paul Offit, I felt his condescension and contempt for white, well-educated parents who research health issues for their children just oozed off the page. The article continues in a bizarre fashion, claiming that those who alter or decline the vaccination schedule are asserting some "special form of privilege" and longs for them to behave like "minority mothers who 'never hesitate' to get vaccines for their children."[22]

Another piece I'd add to the analysis is that doctors are mandated by their profession to push vaccines, and if they deviate from the standard line, they will be investigated. From an article by *California Health Line* in July 2019:

> The California agency that regulates doctors is investigating at least four physicians for issuing questionable medical exemptions to children whose parents did not want them immunized . . .
>
> Vaccine exemptions for medical reasons should be rare, according to the Centers for Disease Control and Prevention. They are typically reserved for children with severely compromised immune systems, like those being treated for cancer or those who are allergic to a vaccine component or have previously had a severe reaction to a vaccine.[23]

These complaints might have a shred of credibility if there was any diagnostic test or set of circumstances which the medical community agreed upon that could serve as the basis for a vaccine exemption.

However, no such test or criteria exists.

Don't believe me?

Ask your family pediatrician for one.

I bet your kids will already be out of the practice before the pediatrician provides you with a single piece of information.

* * *

How crazy are the vaccine mandates getting?

For me, this is personal. For my wife, it may mean she has to leave her job as a speech therapist for Kaiser Health Care in Northern California by September 30, 2021, a profession in which she has worked for the past thirty-three years. And I may have to leave my job as a science teacher for the San Ramon Valley Unified School District, where I've been employed for the past fifteen years. This is from *U.S. News and World Report* on August 5, 2021:

> California will require all of its roughly 2.2 million health care workers and long-term care workers to be fully vaccinated by September 30 as the nation's most populous state is losing ground in the battle against new infections of a more dangerous coronavirus variant.
>
> The order, issued Thursday by the California Department of Public Health, is different than what Democratic Gov. Gavin Newsom said last month when he announced health care workers would have the choice of either getting vaccinated or submitting to weekly testing.
>
> Now, the order does not give health care workers a choice.[24]

There you have it. My wife has already received her letter from Kaiser, and it does not seem that she will qualify for a medical or religious exemption. It appears the concept of "my body, my choice" is officially dead in the Golden State of California. When she asked what she should do, I told her to wait until she gets the official written letter terminating her for refusing a COVID-19 vaccine, because that will be Exhibit #1 in the civil rights lawsuit we may bring against Kaiser.

While it initially seemed that teachers would avoid mandatory vaccination with COVID-19 shots as a condition of employment, that seems to be changing:

> The head of the National Federation of Teachers is calling on educators across the nation to support efforts by some lawmakers to institute mandates requiring teachers to receive the Covid-19 vaccine before returning to in-person class in the fall.
>
> Speaking on NBC's *Meet the Press*, Randi Weingarten reminded viewers that vaccines for other diseases such as polio

are already mandated in schools across the country, while commending teachers who had already chosen to get vaccinated.[25]

What Randi Weingarten neglects to mention is that while there are vaccine mandates for children, to the best of my knowledge, no such mandates exist for teachers in any school district in the United States.

But do we really believe the head of a teacher's union is interested in educating the public, or just frightening us into submission?

How is the idea of a vaccine mandate playing among some traditionally Democratic constituencies? Apparently, not well, if the recent comments of the mayor of Boston are to be believed:

> The mayor of Boston said the city won't be following New York's lead in requiring proof of vaccination at many indoor businesses, claiming the move is reminiscent of "slavery" and birtherism.
>
> Acting Mayor Kim Janney—the first woman and black Bostonian to hold the office—said "there's a long history in this country of people needing to show their papers," when asked Tuesday about the mandate unveiled earlier in the day by Mayor Bill De Blasio that requires proof of vaccination to enter indoor restaurants, entertainment venues and gyms, starting on Sept. 13.[26]

It's going to be interesting to see how the racial angle plays as both the Black and Hispanic communities, as a group, are the most skeptical of all racial groups of COVID-19 vaccines:

> Surveys show that kind of skepticism about the vaccine is widespread. Nearly 40% of Latinos told Pew researchers they would probably or definitely not get the vaccine. More than half of Black respondents said the same.[27]

If such trends hold, it's really not going to be a good look for the Biden administration to throw Hispanics, Blacks, and college-educated white women out of their jobs or prevent them from interacting in society.

However, it's also well-known medical doctors who are commenting on the injustice of these vaccine mandates, such as Dr. Drew Pinsky. A *Breitbart* article from August 4, 2021 reported:

> Dr. David Drew Pinsky, better known as Dr. Drew, tweeted Wednesday that vaccine passports are "modern day segregation."
>
> "The CDC reported today that only 28.4% of the black community have received the Covid vaccine. With states like CA and NY mandating vaccine passports to enter restaurants and businesses, the black community would be most affected by this modern-day segregation," Dr. Drew tweeted.[28]

Aside from sizeable numbers of Hispanics, Blacks, college-educated white women, and medical media luminaries like Dr. Drew, members of law enforcement are also expressing their opposition to mandatory vaccines in, of all places, San Francisco and San Diego, California:

> San Francisco's sheriff's union on Friday threatened that a number of deputies will quit or retire early if they are forced to get a COVID-19 vaccine under the city's mandate.
>
> The policy issued last month requires San Francisco's roughly 35,000 public employees to be vaccinated by Sept. 15 or risk possibly getting fired . . .
>
> The San Diego Deputy Sheriff's Association said on its Facebook page that the ultimatum would force officers from the already short-staffed agency to retire or move on —something it warned the city "cannot afford" . . .
>
> A majority of the Sheriff's Association [San Francisco], which numbers around 700 is vaccinated, the union said. About 160 of its members are not vaccinated, preferring instead to wear masks and test for the virus weekly "due to religious and other beliefs."[29]

The number of vaccine skeptics seems to be growing far beyond just the white, well-educated, Whole Foods moms who like to breastfeed their children and give them food free of pesticides.

* * *

How dangerous are the COVID-19 vaccines?

It's an excellent question and one which every person, regardless of whether you get a COVID-19 vaccine or not, should be interested in. One of the difficulties is that when a new product is introduced into the marketplace, it may take some time for the side-effects to become clear and established. For example, Vioxx was on the market for five years before it was withdrawn because of its side effect of causing heart attacks in otherwise healthy people, with a death toll estimated in the United States as being somewhere between 39,000 to 61,000 deaths.[30]

An excellent article from *Forbes* magazine in 2015 laid out the magnitude of the problem:

> One systematic review that attempted to measure the magnitude of underreporting found that across 37 studies in 12 countries, the median underreporting rate was 94%, according to results published in 2006 in the journal *Drug Safety*. Another review published three years later in the same journal found that under-reporting of adverse events was mainly caused by physician ignorance, the fear of looking ridiculous for reporting merely suspected adverse drug reactions, procrastination, indifference, insecurity and complacency. In a twisted bit of logic, the researchers found that many doctors didn't report adverse events because they figured the FDA only allows safe drugs on the market. Apparently, these physicians fail to realize that regulators need to get their data from somewhere.[31]

You'll forgive me if these findings don't fill me with confidence about the ability of the medical profession to monitor dangerous side effects which may crop up in new medicines or therapies. It also seems that doctors may need some further education regarding their duty to report dangerous side effects.

The monitoring system for vaccines is called the Vaccine Adverse Events Reporting System, or VAERS, for short. It is a voluntary reporting system. No criminal or civil liability attaches to a physician who fails to report an adverse event from a vaccine.

However much longer I may remain a teacher, I do have a duty to report any student I suspect is being abused.

And if I fail to do so, I *do* face criminal and civil liability.

On August 6, 2021, *The Defender*, a publication of Robert F. Kennedy Jr.'s Children's Health Defense organization released their analysis of side effects of the various COVID-19 vaccines from December 14, 2020 to July 30, 2021. They reported:

> Excluding "foreign reports" filed in VAERS, 443,201 adverse events, including 5,739 deaths and 35,881 serious injuries were reported in the U.S.
>
> In the U.S., 344.9 million COVID vaccine doses have been administered as of July 30. This includes 139 million doses of Moderna's vaccine, 193 million doses of Pfizer and 13 million doses of the Johnson and Johnson (J&J) COVID vaccine.
>
> Of the 5,739 U.S. deaths reported as of July 30, 13% occurred with 24 hours of a vaccination, 19% occurred within 48 hours of a vaccination and 34% occurred in people who experienced an onset of symptoms within 48 hours of being vaccinated.[32]

While one might look at those numbers and feel some sense of reassurance, we do not know the extent of underreporting of adverse side effects. If we simply say the amount of underreporting is the same as that of traditional pharmaceutical drugs, that means we have only captured 6 percent of the total and must multiply all those numbers by approximately 15.66 to reach a number which more closely matches reality.

That means instead of 443,201 adverse events, the real number is closer to 6,913, 935.

Instead of 35,581 serious injuries, the real number is closer to 557,198.

That means instead of 5,739 deaths, the real number is closer to 89,528.

The question which nobody can answer definitively is whether the underreporting for vaccine side effects is the same as pharmaceutical drugs, or more or less complete.

My suspicion is that the underreporting of vaccine side effects is greater than the underreporting of traditional pharmaceutical drugs.

* * *

The eugenicists of the 1920s and 1930s were supported by some of the most powerful entities in our country, from the Ivy League universities to the great philanthropic organizations, leaders in the law, the media, and Congress. These groups gave their intellectual and financial support to the evils of the Nazi regime.

That much is beyond doubt.

However, we have a very similar campaign going on today in favor of mandatory vaccination, with many of the same players pulling the strings.

It is also beyond doubt that these eugenicists misrepresented the facts of the *Jacobson v. Massachusetts* case to justify their campaign of approval for involuntary sterilization of those they deemed to be "mental defectives."

These same eugenicists hijacked our government in 1924 and passed the Anti-Immigration Act, essentially barring entry into our country of Italians, Jews, Eastern Europeans, and Asians.

By misrepresenting the facts of the *Jacobson* case, it was an easy slide for the eugenicists to justify involuntary sterilization in this country, block the immigration of those they believed inferior into our country, and give intellectual support to the genocidal racism of the Nazi regime.

We once fought the Nazis and eugenicists with every ounce of our strength. Why are we letting their spiritual descendants get away with the very same thing?

CHAPTER THREE

The Doctors Trial and the Nuremberg Code

Most students of history are aware that after World War II, many of the leading figures of the Nazi regime were put on trial for crimes against humanity by the victorious Allies, in what were later called the "Nuremberg trials," after the city in Germany where the trial was held.

However, what many may not know is that the first group which was put on trial were medical personnel who had aided the Nazi war machine. As described by the United States Holocaust Memorial Museum:

> On December 9, 1946, an American military tribunal opened criminal proceedings against 23 leading German physicians and administrators for their willing participation in war crimes and crimes against humanity. In Nazi Germany, German physicians planned and enacted the Euthanasia Program, the systematic killing of those they deemed "unworthy of life." The victims included people with severe psychiatric, neurological, or physical disabilities. Further, during World War II, German physicians

conducted pseudoscientific medical experiments utilizing thousands of concentration camp prisoners without their consent.[1]

Most are aware of the grisly medical experiments performed on concentration camp prisoners, most notably by Dr. Josef Mengele, often referred to as the "angel of death." But two years before that program got started there was an earlier program which sought to kill German children and adults with disabilities:

> In the spring and summer months of 1939, a number of planners began to organize a secret killing operation targeting disabled children. They were led by Phillip Bouhler, the director of Hitler's private chancellery, and Karl Brandt, Hitler's attending physician.
>
> On August 18, 1939, the Reich Ministry of the Interior circulated a decree requiring all physicians, nurses, and midwives to report newborn infants and children under the age of three who showed signs of severe mental or physical disability.
>
> Beginning in October 1939, public health authorities began to encourage parents of children with disabilities to admit their young children to one of a number of specially designed pediatric clinics throughout Germany and Austria. In reality, the clinics were children's killing wards. There specially recruited medical staff murdered their young charges by lethal overdoses of medication or by starvation.[2]

When one considers what's meant by true evil, nothing fits the bill as much as the killing of children, especially those with disabilities. One can only conclude that this was in essence a test of the German medical establishment. If they would become his willing partner in this Satanic murder of children, Hitler could get them to do anything.

This program continued all through the war years, eventually including children up to the age of seventeen. Conservative estimates suggest that at least ten thousand children were murdered under this program.[3]

* * *

With the success of their euthanasia program for children, the planners quickly drew up plans to kill adult disabled patients living in institutional settings:

> According to Hitler's directive, Fuhrer Chancellery director Phillip Bouhler and physician Karl Brandy led the killing operation. Under their leadership, T4 operatives established six gassing installations for adults as part of the "euthanasia" action. These were:
> - Brandenburg, on the Havel River near Berlin
> - Grafeneck, in southwestern Germany
> - Bernburg, in Saxony
> - Sonnenstein, also in Saxony
> - Hartheim, near Linz on the Danube in Austria
> - Hadamar, in Hessen.[4]

This was without a doubt a massive operation in terms of personnel and materials. This was preceded by a decades-long eugenics campaign, just as Oliver Wendell Holmes, Jr. stewed in his family's cauldron of racial hatred for more than fifty years before finally writing the majority opinion in the horrendous *Buck v. Bell* case. If the media and the government stoke the fires of hatred long enough, you can convince medical people that it's acceptable to murder children.

After that, the murder of adults is almost an afterthought.

The list of conditions for which adults could be subjected to this program was extensive and could easily have been modified to include political opponents. The list included those with schizophrenia, epilepsy, dementia, encephalitis, or other chronic psychiatric or neurological disorders. One could also be subject to the program if you were not of German or "related blood," those adjudged criminally insane, and those confined to an institution for more than five years.[5] This program was administered by a short form which was distributed to all public health officials, private and public hospitals, nursing homes, and mental institutions. As explained by the United States Holocaust Museum:

> Secretly recruited "medical experts," physicians—many of them of significant reputation, worked in teams of three to evaluate the forms. On the basis of their decisions beginning in January 1940, T4 [the name of the program] functionaries began to remove patients selected for the "euthanasia" program from their home institutions. The patients were transported by bus or rail to one of the central gassing installations for killing.
>
> Within hours of their arrivals at such centers, the victims perished in gas chambers. The gas chambers, disguised as shower facilities, used pure, bottle carbon monoxide gas. T4 functionaries burned the bodies in crematoria attached to the gassing facilities.[6]

We like to think of medical professionals as somehow divorced from the passion of their times.

But history, past and present, does not support such an interpretation.

Perhaps because of their intellectual accomplishments, these medical personnel became divorced from reality and could only see the bright, shining future promised by Hitler and not the pile of bodies upon which this new world would be built. In fact, I'd argue that the intellectuals in a society are far more likely to be deceived by such utopian ideals than the common working person, who places more value on the injustice seen today than the nirvana promised for tomorrow.

Ironically, the T4 program to rid Germany of unfit adults quickly attracted attention and knowledge of the program became widespread in Germany. This led to many public and private protests, especially from members of German churches. Probably the most important member of the German clergy who lent his support was the influential bishop of Munster, Clement August Count von Galen. He protested the T4 killings in a sermon delivered on August 3, 1941 which gathered wide attention throughout the Reich.

The T4 program was halted by Hitler in late August 1941, but not before killing an estimated 70,273 individuals between January 1940 and August 1941.

The child euthanasia program continued without interruption during the war.

History seems to suggest it's easier to kill children than adults. Apparently, people notice missing adults before they notice missing children.

* * *

And in the end, what happened in the Nuremberg Trials of the doctors? Was justice served?

> After almost 180 days of proceedings, including the testimony of 85 witnesses and the submission of almost 1,500 documents, the American judges pronounced their verdict on August 20, 1947. Sixteen of the doctors were found guilty. Seven were sentenced to death. They were executed on June 2, 1948.[7]

Ten thousand disabled children were killed and seventy thousand adults. And only seven doctors were sentenced to death. Does it sound to you like a lot of medical personnel escaped justice?

Most of the attention of the Doctors Trial focused on the horrific experiments performed by Nazi doctors in the concentration camps. The Jewish Virtual Library reported:

> There were 70 such "medical research" programs at Nazi concentration camps involving some 7,000 prisoners and some 200

physicians, who worked directly in the concentration camps, but they were not alone. They maintained close professional and research contacts with leading medical institutions and universities and an ongoing relationship with research laboratories. Indeed, the German medical establishment was involved in this work.[8]

It can be clearly seen that the effort to eliminate children and adults with disabilities was of a much grander scale than the experiments conducted in the concentration camps. But the experiments in the camps ignited the most disgust around the world. Some of the more heinous experiments included placing inmates into low pressure chambers, then dissecting their brains while they were still alive; freezing a prisoner, then trying to bring him back to life; inflicting battlefield-like wounds on a prisoner and trying new medications; or artificially inseminating more than three hundred women, then informing them they'd been inseminated with animal, instead of human, sperm and that monsters were growing in their wombs.[9]

All of this horror can be traced in a direct line from the American eugenics program at Harvard, through the mischaracterization of *Jacobsen v. Massachusetts*, to the stunning majority opinion written by Oliver Wendell Holmes Jr. in *Buck v. Bell*.

And what did all of these have in common?

The belief that the state is supreme over the medical decisions of an individual.

* * *

Seven doctors were hung for these crimes.

Dr. Josef Mengele, the "Angel of Death" at Auschwitz Concentration Camp, escaped and apparently lived out the rest of his life in Brazil, supposedly dying there in 1979 of a stroke while swimming.[10]

And what of the others?

> After the war, eugenics was declared a crime against human-
> ity—and act of genocide. Germans were tried and they cited the
> California statutes in their defense. To no avail.
>
> They were found guilty.
>
> However, Mengele's boss, [Otmar Freiherr von] Verschuer
> escaped prosecution. Verschuer re-established his connections
> with California eugenicists who had gone underground and
> renamed their campaign "human genetics." Typical was an
> exchange July 25, 1946 when [Paul] Popenoe wrote Verschuer,
> "It was indeed a pleasure to hear from you again. I have been very
> anxious about my colleagues in Germany . . . I supposed steriliza-
> tion has been discontinued in Germany?" Popenoe offered tidbits
> about various American eugenic luminaries and then sent various
> eugenic publications . . .
>
> . . . Verschuer wrote back, "Your very friendly letter of 7/25 gave
> me a great deal of pleasure and you have my heartfelt thanks for
> it. The letter builds another bridge between your and my scientific
> work; I hope that this bridge will never again collapse but rather
> make possible valuable mutual enrichment and stimulation."[11]

For most of humanity, World War II was an existential fight between
good and evil. For the American and German eugenicists, it was merely
an interruption in their mutual work.

One of the supposedly positive results of the Doctors Trial was
what was called the Nuremberg Code, which was designed to serve as a
template for civilized nations to follow when using medical procedures
which might be called "experimental." While the Nuremberg Code has
never been adopted as law by either the United States or Germany, the
principles in it are supposed to guide medical practice.

Many of those who have objected to mandatory vaccinations cite different parts of the Nuremberg Code as applying directly to the question of mandatory vaccination. From the first section:

> The voluntary consent of the human subject is essential. This means that the person involved should have legal capacity to give consent; should be so situated as to be able to exercise free power of choice, without the intervention of any element of force, fraud, deceit, duress, overreaching, or other ulterior form of constraint or coercion; and should have sufficient knowledge and comprehension of the elements of the subject matter involved as to enable him to make an understanding and enlightened decision.[12]

The idea of "consent" is essential in our business, personal, and medical dealings. We consent to enter into a business relationship, we consent to sexual relations, and we consent to undergo a medical procedure or take a drug. There is probably no idea more central to western civilization than the idea of "consent." Our leaders rule with the "consent" of the governed. When the leaders do not have the "consent" of the governed, then they must leave or they are tyrants. It is a simple and elegant formulation, familiar to each one of us in the very depths of our souls. If we believe each one of us is a unique creation of God, endowed by our Creator with certain inalienable rights, then it follows that we are masters of our body and what goes into it.

And though newspaper columnists are now making the argument that individuals may have the "right" to refuse a vaccine, they do not have a similar right to pursue the occupation of their choice, go to school, or even to venture out into public, thus we are going down a similarly terrifying path.

Those who cite the Nuremberg Code as specifically applicable to the COVID-19 vaccines also point to section three, which states:

The experiment should be so designed and based on the results of animal experimentation and a knowledge of the natural history of the disease or other problem under study that the anticipated results justify the performance of the experiment.[13]

One of the major concerns expressed has been the potential unusual features associated with a coronavirus, which may make it a poor candidate for a vaccine. A coronavirus vaccine was developed in 2004 after the outbreak of SARS-CoV-1 in China and tested on ferrets. This is the abstract from that paper:

Severe acute respiratory syndrome (SARS) caused by a newly identified coronavirus (SARS-CoV) is a serious emerging human disease. In this report, we immunized ferrets (*Mustela putorious furo*) with recombinant vaccine virus Ankara (rMVA) expressing the SARS-CoV spike (S) protein. Immunized ferrets developed a more rapid and vigorous neutralizing antibody response than control animals after challenge with SARS-CoV; however, they also exhibited strong inflammatory responses in liver tissue. Inflammation in control animals exposed to SARS-CoV was relatively mild. Thus, our data suggests that vaccination with RMVA expressing SARS-CoV S [spike] protein is associated with enhanced hepatitis.[14]

In essence, with the vaccine for the first SARS-CoV virus they were using a spike protein to confer immunity, just as we are doing today. At first the experiment seemed to be a success as antibodies developed, but when there was an exposure to SARS-CoV-1, the immune system overreacted, causing inflammation and damage in the liver tissue. This is a condition known as "antibody dependent enhancement," which involves your immune system overreacting to a virus or pathogen.

Your body's overreaction is what causes the harm, not the virus.

In plain English, the ferrets would have been better off *not* getting a vaccine and letting their immune system handle the virus on its own.

While it is impossible for anybody to gauge the nature of this threat, *Nature Microbiology* published an article on antibody-dependent enhancement of any possible COVID-19 vaccines in October 2020 and stated there is a great deal about these risks we do not know. They wrote:

> Antibody-based drugs against severe acute respiratory syndrome coronavirus 2 (SARS-CoV-2) are being expedited through preclinical and clinical development. Data from the study of SARS-CoV-2 and other respiratory viruses suggest that anti-SARS-CoV-2 antibodies could exacerbate COVID-19 through antibody-dependent enhancement (ADE). Previous respiratory syncital virus and dengue virus vaccine studies revealed human clinical safety risks related to ADE, resulting in failed vaccine trials.[15]

The leading journals in medicine are openly discussing the problem of antibody-dependent enhancement. One can understand the bind in which many medical professionals find themselves. They know that infection with SARS-CoV-2 may cause a dangerous inflammatory response, but the vaccine also has a similar risk.

It is in exactly these situations that the truth is most important, and the ability to make your own decisions about your healthcare is so vital. Researchers from the Massachusetts Institute of Technology (MIT) reviewed the available literature and discussed some of their concerns in a May 10, 2021 publication:

> We then review both components of and the intended biological response to these vaccines, including production of the spike protein itself, and their potential relationship to a wide range of both

acute and long-term induced pathologies, such as blood disorders, neurodegenerative diseases and autoimmune diseases. Among these potential induced pathologies, we discuss the relevance of prion-protein-related amino acid sequences within the spike protein. We also present a brief review of studies supporting the potential for spike protein "shedding", transmission of the protein from a vaccinated to an unvaccinated person, resulting in symptoms induced in the latter.

We finish by addressing a common point of debate, whether or not these vaccines could modify the DNA of those receiving the vaccination. While there are no studies demonstrating definitively that this is happening, we provide a plausible scenario, supported by previously established pathways for transformation and transportation of genetic material, whereby injected mRNA could ultimately be incorporated into germ cell DNA for transgenerational transmission.[16]

The overwhelming message from these researchers is that we have no idea of the long-term consequences of messenger RNA vaccines, which is less of a traditional vaccine, and more of a gene therapy.

There is no other way of saying it.

The COVID-19 vaccines, rushed through at "warp speed," are experimental therapies, and should be covered under the principles of the Nuremberg Code of informed consent and understanding the potential benefits and risks of treatment.

* * *

In an era where we are gaining unprecedented freedoms, such as the ability to use cannabis for medical or recreational purposes, you may not have the choice for what medications to put in your body, especially if you are unconscious. A 2012 article in the magazine *New Scientist* by author Harriet Washington looked at how the principles of the Nuremberg Code had been repeatedly violated by corporations and governments:

> My book, *Medical Apartheid* documents many cases. In 1994, for example, the Medical University of South Carolina in Charleston was accused of enrolling poor black women into narcotic-treatment research without their knowledge. The next year in Los Angeles, an experimental measles vaccine was tested on children, mostly black and Hispanic, without their parents' consent. In 1994 and 1995, New York City law enforcement officials helped researchers coerce black parents into enrolling their boys into a study that sought to establish a genetic propensity for violence, again without their consent. And in 2001, the Kennedy-Kriger Institute in Baltimore was found guilty by a Maryland court of encouraging black families to move into lead-contaminated houses as part of a study on lead levels in children (the verdict was later overturned.)[17]

It's genuinely shocking to read about these examples from the 1990s and early 2000s, more than fifty years after the Nuremberg Trials. Many in the vaccine-skeptical community talk of their suspicions that the real agenda behind mandatory vaccination is more in line with the eugenics agenda of American and German scientists of the 1930s and 1940s.

As an attorney, I've learned that the same evidence can generate multiple scenarios which are equally plausible. I believe it is beyond question that American eugenicists, beginning in 1875 with Oliver Wendell Holmes Sr., gave an intellectual framework as well as significant financial support to Hitler's genocidal racism.

I believe it's equally clear that after World War II, many of these same American and German eugenicists escaped punishment for their actions, either in criminal courts or in public shame.

The legal framework for their actions was established by the misinterpretation of *Jacobson v. Massachusetts* in 1905 and the majority opinion by Oliver Wendell Holmes Jr. in *Buck v. Bell* in 1927, which allowed the state to involuntarily sterilize a psychiatric patient. With those legal precedents established, it was a short hop to the euthanasia programs of Hitler's Germany, as well as the horrifying medical experiments performed in the concentration camps.

From my reading of history, the trail of any remaining eugenicists goes cold after World War II.

There are two possible explanations.

The eugenicists could have decided to continue their work in secret by hiding in plain sight and simply changing the name of their programs to something more palatable.

Or they could have abandoned their ideology.

I am unable to decide between these two choices.

What I can say with certainty, though, is that those who remained in the upper echelons of public health didn't want to let go of the tools of government control brought into being through the actions of the eugenicists. Once the government or industry can perform experiments on vulnerable minorities without risk of punishment, it gives those same entities license to expand their reach:

> Since the 1980s, around 20 US research projects have won legal waivers that allow them to bypass any form of consent. From 1990 until 2005, for example, the Department of Defense obtained a waiver that allowed it to force 8.9 million ground troops to accept inoculation with experimental anthrax vaccines.
>
> Civilians' rights are violated, too. In 1996, U.S. law was changed to permit non-consensual research on trauma victims on the pretext they are unconscious and unable to give their consent . . .
>
> . . . In 2003, biotechnology company Northfield Laboratories set up a nationwide trial of its blood substitute Poly Heme. The

substance was randomly administered by ambulance crews to unconscious victims of car accidents, shootings and cardiac arrests.[18]

Because of the misinterpretation of *Jacobson v. Massachusetts*, the government has grabbed the authority to perform medical experiments on members of the armed forces, as well as unconscious citizens.

However, if you can't experiment on your own troops or unconscious Americans, you might be able to outsource your research:

> In 2011, drug giant Pfizer paid $75 million to settle claims that children in Kano State, Nigeria were injured or killed by non-consensual administration of its experimental meningitis drug Trovan. Just as U.S, physicians demanded justice at Nuremberg, Nigerian parents stormed courts in Kano and Manhattan to demand that we live up to our stated ideals. Sixty-five years on, it is high time we did.[19]

As I write these words in 2021, we are nearly seventy-five years removed from the promulgation of the Nuremberg Code. And yet the principle of informed consent, as well as the ability to make such decisions free of pressure, such as losing your job, not being able to go to school, or not being able to participate in public life, remains under threat.

In an age where we can smoke pot without fear of arrest, it seems inconceivable that we have lost the right to decide what to put in our bodies.

It is time to fight to regain our rights from those who would seek to take them from us.

We have been down that road before and we know it leads to a place of unspeakable evil.

CHAPTER FOUR

The 1986 National Childhood Vaccine Injury Act

If I turn on the television today I might see ads for lawyers telling me I can sue the Catholic church if I'd been sexually abused by their clergy; I can sue Monsanto if I developed non-Hodgkin's lymphoma from using Roundup; or if I developed Parkinson's disease I can sue the manufacturer of an herbicide called Paraquat.

But who do you sue if you have a vaccine injury?

Many readers may be puzzling this question in their head because they'd never even considered the possibility. I know because I've asked more than a few of my really smart friends, only to have them look as if I urged them to swallow a goldfish.

I assure you it's not a trick question.

Like Sherlock Holmes, sometimes the most important clue, such as in the case of the dog that didn't bark, is the absence of evidence.

What should be present, but is missing?

I have diligently investigated this question, putting most of what I found into a book called *Inoculated: How Science Lost Its Soul in Autism*. While I certainly believe there are villains in this story, the most remarkable revelation I came to was that a system had been set up to discourage any questions about vaccines and their potential side effects.

And I also found some surprising allies, who were raising many of the same questions I had.

* * *

The simple fact of the matter is you *cannot* sue a vaccine manufacturer if you suffer a vaccine reaction, nor can your heirs, if that reaction is so severe it kills you.

Let me talk about how a typical claim is brought if a product you bought causes injury, and how vaccines have a special carve-out in the law.

Imagine you buy a toaster because you like to have a nice, toasted bagel with cream cheese in the morning before you go to work. However, instead of turning that bagel a nice, crispy brown, it catches on fire, igniting the drapes, and burns down your kitchen. In addition, in the midst of fighting the blaze, you suffer some smoke inhalation.

You find yourself a good products liability lawyer; he sues the company which made the toaster; and in a process called "discovery" the toaster company has to turn over all the evidence they have of complaints from customers about their toaster catching fire, as well as any internal documents from researchers or executives about the propensity of that toaster to catch fire.

In figuring out if your toaster was the first to ever catch fire, or was case number 725, you'll know whether you have a good case.

The idea is that both the manufacturer and the plaintiff should have the same information on the safety of that product.

Nothing even remotely like that exists regarding vaccines. Instead, the company can retain whatever records it wants, not share them with a plaintiff, and instead of facing off against corporate attorneys, you will find yourself up against lawyers from the US Department of Justice.

That's right. If you sue for a vaccine injury, you will face the full might of the United States government.

You probably want to know how that happened.

* * *

One individual with whom I conducted a long interview for my book *Inoculated* was Stanford University law professor Nora Freeman Engstrom. She is one of the few academics who has studied the working of the Vaccine Court, or as it is properly known, the "Vaccine Injury Compensation Program," or VICP.

Before our interview I had reviewed her eighty-eight-page law review article from the *University of Pennsylvania Law Review* in 2015 titled "A Dose of Reality for Specialized Courts: Lessons from the VICP." In her discussion of how vaccines were moved from the traditional civil court system into this new specialized court, she wrote:

> These side effects received little attention until the 1970s when three unrelated events seemed to conspire to unsettle the status quo. The first was the "notably troublesome" case of Anita Reyes. While an infant living near the Mexican border, Reyes received a dose of Wyeth's polio vaccine and, shortly thereafter, contracted polio, resulting in her near-complete paralysis. Her parents initiated suit on her behalf, a jury ruled in her favor, and, in 1974, the Fifth Circuit affirmed. Both the Fifth Circuit's holding and its reasoning were, for vaccine companies, ominous. Reyes v. Wyeth, that is, substantially expanded liability for vaccine manufacturers and also explicitly invoked enterprise liability principles to justify that expansion – suggesting that, going forward, between victims and vaccine manufacturers, the latter should bear the loss.[1]

The result seemed to be in accordance with what most of us would think of as justice. A company makes a vaccine, they give it to a young girl, she develops polio which is so severe that it results in "near-complete paralysis" for her, and the company pays for the harm it caused. If this was the only time this vaccine caused polio, maybe it was just a fluke. However, if this is happening a lot, you should probably withdraw the product until you better understand what's going on.

Next up was the swine flu vaccine catastrophe, which many commentators talk about as though it never took place:

> Then, on the heels of the Reyes decision came the swine flu faux epidemic of 1976. There, forty-five million Americans (one third of the adult population) subjected themselves to a flu shot at President Gerald Ford's stern urging only to learn, later, that the flu was not particularly dangerous, but the shot itself was—causing in some small proportion of patients Guillain-Barré syndrome, a usually reversible but occasionally fatal form of paralysis. A flood of litigation and withering press attention followed.[2]

Swine flu was predicted to devastate the population, and yet it did not. Instead, it caused a terrible neurological disease. This is an episode we should remember as there is this massive stampede toward getting a vaccine which was developed at "warp speed."

The third strike against vaccines was the use of a dangerous DTP vaccine, which threatened to destroy the vaccine industry:

> Rounding out this troubling trilogy, on April 19, 1982, an NBC affiliate aired an Emmy-winning, hour-long television documentary titled DPT: Vaccine Roulette. With footage of dead infants and convulsing children, the broadcast charged that the pertussis component in the DTP vaccine (which protected against whooping cough) had not been adequately tested and could cause "damage to a devastating degree."[3]

This vaccine killed children, or caused seizures. That was the accepted reality of vaccines in 1982. There was none of this foolishness that vaccines don't cause injury, or that it's so rare that it might be just one in a million cases. In fact, the potential damages for harm to children by this single vaccine threatened to put many companies out of business:

> As the number of lawsuits ticked upward, so did manufacturers' dismay. In 1984, for example, Lederle's President went on record declaring "[t]he present dollar demand of DTP lawsuits against Lederle is 200 times greater than our total sale of DTP vaccine in 1983" . . . Another vaccine manufacturer—Connaught Laboratories—faced a similar plight, as suits filed against it in 1985 and 1986 sought a combined billion dollars in damages.[4]

The damages from just a single vaccine threatened to wipe out the entire industry. So, what was the response? Spend a lot of money on vaccine safety so they could better understand the biological interactions and prevent children from being harmed?

No.

The answer was to indemnify the vaccine companies by putting the US government on the hook, and drastically scale up the childhood vaccine program. It's as if there was a single drunk driver on the road, and the resolution for the problem was to get a lot more cars and a lot more alcohol.

The pharmaceutical companies threatened to pull out of the childhood vaccine market, and the politicians responded, led by Congressman Henry A. Waxman of California. Their answer was the 1986 National Childhood Vaccine Injury Act, which President Reagan signed into law, despite strong reservations. The *New York Times* reported:

> Mr. Reagan said he had approved the bill "with mixed feelings" because he had "serious reservations" about the vaccine compensation

program . . . The program would "be administered not by the executive branch, but by the Federal judiciary," Mr. Reagan said, calling it an "unprecedented arrangement" that was inconsistent with the constitutional arrangement for separation of powers among the branches of the Federal Government.[5]

The public health officials had everything they needed now. They could claim legal authority to force vaccinations on children and their Big Pharma partners were off the hook for any injury or deaths they caused to children.

And nobody imagined the catastrophe to come.

* * *

In my interview with Stanford Law School Professor Nora Freeman Engstrom, we covered many issues, not the least of which was how this "no-fault" system, which was supposed to speed recovery to the families of vaccine-injured children, was neither quick nor satisfying to the parents who entered the program. At one point, I asked her whether vaccine safety would have been better handled under the traditional civil court system. She seemed conflicted by the question and said:

> Vaccine injuries are super hard. Let's not forget how hard it is to get compensation for your vaccine-injured kid in court. You've got a whole lot of things working against you. You've got the fact that vaccines tend not to cause signature diseases. And so the causation questions are always going to be hard. You've got to show deviation from custom, and that gives doctors tons of latitude. You have to show the product is defective to argue against the manufacture. You have to identify the manufacturer. You have to do all of this in the statute of limitations. It's really, really hard. And you have to find a lawyer who's willing to do this on a contingency basis when fewer lawyers are willing to do it.[6]

As one of the few academics to study the workings of the vaccine court, Engstrom's opinion should be among the most credible in any debate about vaccines. The approach of the mainstream media, as best I've been able to determine, is to act as if there's no such thing as a special court where claims of vaccine injury are adjudicated, or that their rules of evidence are biased in favor of the manufacturer.

When a Stanford law professor tells you that proving vaccine injuries is "super hard," because they "tend not to cause signature diseases," we have some inkling of how incomplete the coverage of this issue has been in the mainstream media.

I also discussed with Professor Engstrom the proposals made by others, such as enlarging the ability of patients to obtain internal documents from vaccine makers, and lengthening the statute of limitations, as has been done most recently in sexual abuse cases involving the Catholic Church, or other large organizations. She replied:

> I am totally sympathetic to that view. It sounds right to me. As an academic, I think there is something so interesting about the VICP [Vaccine Injury Compensation Program]. Some things are designed to fail. Auto no-fault in most states was designed to fail. It was a compromise and the trial lawyers' associations were able to embed poison pills into the legislation. So everybody knew it wasn't going to work out well, and it didn't surprise anybody. And you can say auto no-fault wasn't successful and that doesn't say anything bad about the no-fault idea, because of the way it had poison pills in it. The VICP is so interesting because it was designed to succeed. On the face of it, it should have succeeded.[7]

Nevertheless, the Vaccine Injury Compensation Program has been a failure, at least in the eyes of some. Normally when something so vital to the public's health is questioned, we get extensive coverage of it in the news. But vaccines have escaped any of that scrutiny. The average person does not even formulate the question of whether our strategy for getting to the truth about vaccines is even remotely similar to the way the public gets the truth about all the other consumer products in their lives.

It is time for people to start formulating these questions, asking them in public, and pressing for the answers.

*   *   *

In my investigation of the Vaccine Court, I also conducted a long interview with Gary Golkiewicz, who joined the court on its creation on October 1, 1988 and served as its Chief Special Master (roughly equivalent to a Chief Justice) until 2010. I began by asking him to explain the initial idea of the court:

> They developed the program for the DPT vaccine. And the story I heard is it all began with a pediatrician who gave a DPT to a friend's son and the son suffered a reaction. And the friend did not want to sue the doctor because they were friends. They basically settled with whatever insurance coverage the doctor had. And the doctor thought that was grossly unfair. And that was the genesis of the idea. A lot of people are under the impression that the manufacturers were the big pushers of the legislation. But the biggest backer of the program was the American Academy of Pediatrics. Through the years the manufacturers have been very careful politically, in interacting with me and the court.[8]

Whenever one begins an investigation you find surprises along the way. I had never even considered the idea that the American Academy of Pediatrics would have been the main backer of the legislation. My money would have been on the pharmaceutical companies, but in reality, it seemed that they took a wait and see approach.

One of the claims I have heard people make is that the Act was designed to cut off frivolous claims by the parents, but former Chief Special Master Golkiewicz said that wasn't the case. The Act was designed to speed compensation to the families of those injured by vaccines so they could quickly continue with their lives. However, there were immediate benefits to the manufacturers:

> One would say the Act supports to manufacturers as well because they wanted to cut off litigation, which the Hill wanted as well. The Act was designed to cut down financial pressure on the manufacturers, to create an environment for the research and development of new vaccines. To do that, they created the table that was over-inclusive. And one would argue that the manufacturers were very supportive of that, because it has done exactly what it was designed to do, and that is to cut off litigation against the manufacturers.[9]

According to the former Chief Special Master, although the manufacturers were not in charge of the effort to remove liability from childhood vaccines, they certainly enjoyed the benefit. However, while at first many argued the table of injuries was generous, in 1995 it was drastically reduced, leading to many injuries no longer being compensated.

One of the claims I'd heard in the autism community was that there was evidence that at least eighty-three children had been compensated for their autism caused by vaccines. I questioned Golkiewicz directly about this accusation and this was his answer:

> Before the autism cases were filed [a class action known as the Autism Omnibus Proceeding], we had compensated cases in which a child suffered an encephalopathy, and then went on to exhibit symptoms of autism, or had autism. That was a table case. If you got the covered vaccine, got the encephalopathy within three days of a vaccine, you were compensated. Now where the

autism came in was not on the causation side of the equation.
Autism came in on the damages side, in determining what the
lifetime needs of the child were. So long before the autism cases
were filed, autism cases were compensated, but they were compensated as a byproduct of the encephalopathy.[10]

It's striking how this is the perspective of the guy who ran the Vaccine
Court for more than two decades, but you'd be hard-pressed to find
anything remotely similar in any mainstream media article.

This is how I interpret his remarks. The vaccines can cause encephalopathy (swelling or inflammation of the brain), and one of the damages caused by such inflammation is autism. In my mind, it's A causes
B, which in turn causes C. Therefore, I believe it's accurate to say
vaccines cause autism. If you want to be precise, you might add it's
because of the brain swelling caused by the vaccines. Special Master
Golkiewicz even detailed how he would go about determining whether
a child's autism was caused by the vaccine:

> What I'm looking for is the diagnosis of what this individual has.
> I was told early on that the record you should go to is the first
> neurology visit. The parents are going to give a history, in the last
> three or five days my child started doing this, and you may even
> see a vaccination as part of that history. But you can go back and
> find when the vaccination took place. The parents can say these
> are the events that took place, and then in the differential by the
> doctor, there might be encephalopathy. So you piece it together.
> You very seldom find one record that is a smoking gun.[11]

When the approach was explained to me in this manner it seemed
clear Golkiewicz was attempting an honest effort to get to the bottom
of whether a vaccine had devastated a child's future, and if so, how
much compensation should be paid to the parents for a lifetime of
taking care of that child. As far as the problems, which have developed

with the program since the tightening of the table of injuries in 1995, Golkiewicz said, "You could actually write that it's fallen into the same black hole that we argue about with other government programs, like the milk subsidy, or the sugar subsidy. Once they get on the books, they stay on the books."[12]

My interview was conducted with Golkiewicz in 2016 and this is what he believed at the time about the troubles with the program, and how they might possibly be addressed:

> We're operating today with an Act designed to handle the DTP shortage, which we no longer even give. The original Act had I believe 6 vaccines, but now I think we're up to 18 or 19. So that original language is being applied to the HPV [Human Papilloma Virus] vaccine and also the flu vaccine, which is now the number one source for work in the program. And it's not effective. And that's what's causing all the frustration from the parent's side, which is absolutely correct. The argument that the program takes too long, it's too litigious, and it's not quick justice. They expected a hundred and fifty cases a year and last year they hit over nine hundred. And this year they're on track to go over a thousand cases.[13]

On a personal note, I have to say it was quite refreshing to talk to the former Chief Special Master of the Vaccine Court as a fellow attorney. For those of us with a legal background, we understand the importance of being able to see both sides of an issue. Probably every person with a law degree today has had the experience of being in a law school classroom and having the professor ask you to present one side of an issue, then ask you to flip sides and make the best argument for the other side. It promotes a certain flexibility of mind, which I fear many of us are losing today. In many ways, we are essentially becoming blind to the valid arguments of those with whom we disagree.

It was also clear from my interview that Golkiewicz had tried to avoid the catastrophe which has befallen the Vaccine Court since 1995:

> I told people in 1995 when they were tightening down this table what the impact was going to be. And it did at that time. But it got exacerbated by the addition of all these new vaccines with no table of events, including the flu vaccine. The number one vaccine and injury right now is the flu vaccine with Guillain-Barré syndrome following the vaccine. So the number one injury that's being compensated in the program has no table.
>
> Every problem that comes up or is talked about in the vaccine program, I can trace back to the table versus causation in fact. If you went in and could add a table of events for each vaccine, a generous table, as they talked about in the legislative history, you would not have the issues you are talking about now.[14]

In the most delicate part of my interview, I asked the former Chief Special Master about some of the allegations of vaccine whistleblowers, such as the Simpsonwood Conference of June 2000, and the allegations of senior CDC scientist Dr. William Thompson from 2014. Since that time, another whistleblower has come forward, the government's own chief witness in the Autism Omnibus Proceeding, Dr. Andrew Zimmerman of Johns Hopkins University. (I will be covering each of these situations in later chapters.)

I understood I was putting the former Chief Special Master in something of an awkward position, as a magistrate has no authority to investigate claims on their own and can only rule on the evidence presented in their court. To his credit, Golkiewicz did not shy away from the question:

> I would say this in answer to your question. The Special Masters, like any judge, rely upon information. If that information is not correct, it would obviously impact the information the judge is

considering and could potentially impact his decision. In that respect, the Special Masters are no different than any other judge. Your decision is only as good as the information you get.[15]

The vaccine issue shouldn't be different than any other issue. We should be asking the question of whether we have a robust system to determine the truth of their safety. At the very least, since we are giving these treatments to children, the system for determining their safety should be at least as strong as that used for our faulty toasters.

You might think to yourself I've made a good theoretical argument about why vaccines shouldn't be protected by a special court which functions as a de facto arm of the federal government. But what do the numbers actually show? How much has been paid out for vaccine damages?

According to a March 25, 2021 article from CNBC, "The VICP has paid out approximately $4.5 billion dollars in total compensation as of March 1, 2021."[16]

If we return to the 2006 study in the journal *Drug Safety*, which estimated only 6 percent of negative drug side effects were reported,[17] that $4.5 billion dollars would balloon to about $70.5 billion dollars.

However, in 1993, FDA Commissioner David Kessler estimated that only about 1 percent of serious adverse events to prescription drugs were reported.[18] If that was an accurate estimate of the under-reporting of vaccine injuries, the actual damage would be around $445.5 billion dollars.

I suspect the under-reporting of vaccine injuries to be much greater than that for prescription drugs, and thus, the damage to be even greater than the estimates I've already provided.

Near the end of each of my interviews with Stanford law professor Nora Freeman Engstrom and former Chief Special Master Gary Golkiewicz, I asked if in their professional opinion they thought the so-called "Vaccine Court" had protected the public.

Both declined to answer my question as being beyond their ability to provide an informed opinion. This struck me as shocking. How is it that a Stanford law professor and the man who ran the Vaccine Injury Compensation Program for more than twenty years can't even venture an opinion as to whether the public is being protected?

I felt as if I was interviewing generals of the Vietnam War, honest and ethical public servants, who could provide me with all the data I needed but could render no ultimate judgment as to whether the effort should have ever been attempted. How is it that these leading figures in our society can offer no insight as to whether we have made a dreadful mistake for which future generations will rightfully condemn us?

It is up to us to get answers from our leaders.

* * *

One would expect that when a law is named the National Childhood Vaccine Injury Act, it would apply only to children.

However, you would be wrong.

In October 2010, the US Supreme Court took up the case of *Bruesewtiz v. Wyeth Labs*. On February 22, 2011, in a 6–2 decision, the Supreme Court ruled that the National Childhood Vaccine Injury Act of 1986 applied to adult vaccines as well.

That's right: a conservative Supreme Court ruled that "childhood" meant adults as well. So much for the traditional meaning of words.

It was left to the two most liberal justices on the Supreme Court at that time, Sonia Sotomayor and Ruth Bader Ginsburg, to point out the absurdity of this ruling. Sotomayor wrote a blistering twenty-eight-page dissent, a portion of which read:

> Vaccine manufacturers have long been subject to a legal duty, rooted in basic principles of products liability law, to improve the designs of their vaccines in light of advances in science and technology. Until today, that duty was enforceable through traditional state law tort action for defective deign. In holding that

section 22(b)1) of the National Childhood Vaccine Injury Act of 1986 (Vaccine Act or Act), 42 U.S.C. section 300aa22(b)(1), pre-empts all design defect claims for injuries stemming from vaccines covered under the Act, the Court imposes its own bare policy preferences over the considered judgment of Congress. In doing so, the Court exercises 13 words from the statutory text, misconstrues the Act's legislative history, and disturbs the careful balance Congress struck between compensating vaccine-injured children and stabilizing the childhood vaccine market. Its decision leaves a regulatory vacuum in which no one ensures that vaccine manufacturers adequately take account of scientific and technological advancements when designing or distributing their products. Because nothing in the text, structure, or legislative history of the Vaccine Act remotely suggests that Congress intended such a result, I respectfully dissent.[19]

A supposedly conservative Supreme Court had just made up a new law. Now, *all* vaccines were subject to the Vaccine Court and plaintiffs had to face off against a battalion of lawyers from the US Justice Department. The pharmaceutical companies had more than twenty years of immunity from claims involving devastating childhood vaccine injuries and were preparing to have immunity for the adult market as well.

It had been quite a reversal from the case of *Reyes v. Wyeth Laboratories* in 1974, which placed the liability for vaccine injuries upon vaccine manufacturers. Thirty-six years later, in 2010, *Bruesewitz v. Wyeth Laboratories* removed all the remaining adult vaccines out of the traditional civil court system and placed them inside the poorly functioning Vaccine Court.

There has probably not been a greater miscarriage of justice since the *Buck v. Bell* case in 1927, which found that being sterilized against your will was no different than the five dollar fine imposed in the *Jacobson v. Massachusetts* case of 1905.

Any hope the public might have had that the legal profession might protect them had vanished with the *Bruesewitz* case. Establishing the safety of vaccines would now be left solely in the hands of scientists and public health officials.

A few brave scientists would try to stand up against this onslaught. But the financial and organizational juggernaut of money, power, and prestige would ensure their voices would not be heard.

CHAPTER FIVE

The Simpsonwood Conference (2000) — Vaccine Contempt Becomes Vaccine Conspiracy

On June 7 and 8, 2000, a very unusual conference was held at the Simpsonwood Retreat Center in Norcross Georgia.

There were fifty-three individuals at this meeting. Fifteen were employees of the Centers for Disease Control and Prevention. Fourteen individuals had academic appointments which overlapped with other affiliations. Twelve individuals directly represented the National Immunization Program. Five were representatives of the vaccine companies, SmithKline Beecham, Merck, Wyeth, North American Vaccine, and Aventis Pasteur. There were four representatives of the American Academy of Pediatrics. Three individuals represented the Food and Drug Administration. Three individuals were not medical doctors, but a project manager for the Northern California Kaiser Vaccine Study Center, the data manager for the same Kaiser group, and a "health communication specialist" for the National Immunization Program. There were no patient advocacy groups or outside watchdog groups invited.

What was their purpose in meeting? They were there to discuss how to handle a very disturbing signal from the data that the increasing number of vaccines was leading to increasing numbers of developmental problems in children. They would argue whether it was the mercury, or possibly the aluminum in the vaccines, or something else they hadn't even considered.

But while the discussion among the medical and scientific professionals was frank and open, it would be nothing like the messaging they would give the public in the years to come. The message the public would get was that vaccines were "safe and effective" and that the chances of an adverse reaction were "a million to one."

* * *

The assembled guests were welcomed by Dr. Walter Orenstein, the director of the National Immunization Program at the Centers for Disease Control.

> I want to thank all of you for coming here and taking time out of your very busy schedules to spend the next day and a half with us. Not only do we thank you for taking the time, but for taking the time on such short notice, and also putting up with what I gather those of us who aren't townies here didn't realize. But apparently, the biggest meeting in Atlanta, which has taken up all the hotel space and all of the cars. So I think many of you had to take taxis here. We appreciate you putting up with this.
>
> But at least we did arrange the weather nicely and you can look out occasionally and see some beautiful trees.
>
> I think I am particularly impressed with the quality of expertise. We truly have been able to get, at very short notice, some of the most outstanding leaders in multiple fields. That will be important in interpreting the data.
>
> We who work with vaccines take safety very seriously. Vaccines are generally given to healthy children and I think the public has, deservedly so, very high expectations for vaccine safety, as well as

the effectiveness of vaccination programs. Those who don't know, initial concerns were raised last summer that mercury, as methylmercury in vaccines, might exceed safe levels. As a result of these concerns, CDC undertook, in collaboration with investigators in the Vaccine Safety Datalink, an effort to evaluate whether there were any health risks from mercury in any of these vaccines.[1]

When you read an opening like that from the director of the National Immunization Program, you realize it's an "all-hands on deck" type of situation. As Orenstein moved further into his remarks, he noted there were concerns about a "possible dose response effect of increasing levels of methylmercury and certain neurologic diseases," and noted that the meeting was not for policymaking but should be considered "an individual simultaneous consultation."[2]

The first presentation was by Dr. Roger Bernier, the associate director for science in the National Immunization Program at the CDC, and he reviewed the history of the mercury derivative, thimerosal, which was used in vaccines at that time.

Next to present was Dr. Dick Johnston, an immunologist and pediatrician at the University of Colorado School of Medicine, and he went deeper into the explanation of thimerosal:

> Thimerosal functions as an anti-microbial after it is cleaved into ethylmercury and thiosalicylate, which is inactive. It is the ethylmercury which is bactericidal at acidic PH and fungistatic at neutral and alkaline PH. It has no activity against spore forming organisms. There is very limited pharmokinetic data concerning ethylmercury. There is very limited data on its blood levels. There is no data on its excretion. It is recognized to both cross placenta and the blood-brain barrier. The data on its toxicity, ethyl-mercury, is spare. It is primarily recognized as a cause of hypersensitivity. Acutely it can cause neurologic and renal toxicity, including death from overdose.[3]

Let me break down some of the science for you. When they say there is "limited pharmokinetic data" that means they don't know how the body responds to it, especially in the presence of other medical interventions, such as antibiotics, or even a Tylenol you might give a baby when it's teething.

To round out this horror show, let's talk about all the other things they don't know. They don't know about the mercury levels in the blood. They don't know how quickly mercury gets excreted. They do know mercury crosses the placenta and the blood-brain barrier. That means it gets to your baby and your brain. The data on toxicity is thin. They only identify overexposure when you get sick. And finally, it can cause neurologic and renal problems, in addition to death.

In one of the most terrifying passages from the transcript, Dr. Johnston then talked about an enormous blind spot among those who study vaccines:

> As an aside, we found a cultural difference between vaccinologists and environmental health people in that many of us in the vaccine arena had never thought about uncertainty factors before. We tend to be relatively concrete in our thinking. Probably one of the big cultural events in that meeting, at least for me, was when Dr. Clarkson repetitively pointed out that we just didn't get it about uncertainty. And he was actually quite right. It took us a couple of days to understand the factor of uncertainty in assessing environmental exposure, especially to metals.[4]

Top scientists involved in public health have trouble with the concept of "uncertainty." In fact, it had to be pointed out to them "repetitively." Please feel free to insert your own joke about doctors believing they are God. These are the people making health decisions for hundreds of millions of people and they have difficulty grasping the concept they may be wrong. Sometimes reality is far more unbelievable than any fiction.

However, to his credit, Dr. Johnston seems to have finally grasped the concept of "uncertainty" and realized there was more to the vaccine issue than simply mercury. There was also the aluminum problem:

> Finally, I would like to mention one more issue. As you know, the National Vaccine Program Office has sponsored two conferences on metals and vaccines. I have just recounted a summary of the mercury, the thimerosal in vaccines. We just recently had another meeting that some of you were able to attend, dealing with aluminum in vaccines. I would just like to say one or two words about that before I conclude.
>
> We learned at the meeting a number of important things about aluminum, and I think they are important in our consideration today.
>
> First, aluminum salts, and there are a number of different salts that are utilized, reduce the amount of antigen and the number of injections required for primary immunization. Secondly, they don't have much role in recall immunization, but it would represent a significant burden to try and develop different vaccines for primary and subsequent immunization. Aluminum salts are important in the formulating process of vaccines, both in antigen stabilization and absorption of endotoxin. Aluminum salts have a very wide margin of safety.
>
> Aluminum and mercury are often simultaneously administered to infants, both at the same site and at different sites. However, we also learned that there is absolutely no data, including animal data, about the potential for synergy, additivity or antagonism, all of which can occur in binary metal mixtures that relate to and allow us to draw any conclusions from the simultaneous exposure to these two salts in vaccines.[5]

The readers of this book are probably hearing about the aluminum for the very first time. And yet, it is an enormous concern. While aluminum is an abundant element in the Earth's crust, it is only in the last hundred years or so that we've been able to chemically change it into various aluminum salts and use them in a wide variety of medical and other consumer products. However, you can ask any biologist of the importance of aluminum to the body, and they will tell you that aluminum has *no* biological importance to the health of an organism. In other words, since life on Earth has *never* had to process aluminum, it has never had any need to deal with it, except for the last decades when we have used this chemically altered form of aluminum in our products. The ubiquitous use of aluminum in our world today poses a unique challenge, and as with mercury, we have little data on its potential effects.

And even if mercury has been removed from the vaccines as claimed, there is still an enormous environmental burden of mercury in our environment, and it is present in varying degrees in all our bodies.

Thus, we do not know how the injected aluminum salts in vaccines, or in other medications called "biologics," might react with the mercury already present in our bodies.

There was then a short talk by Dr. Frank DeStefano (a major player in the case of whistleblower Dr. William Thompson, covered in the next chapter) followed by the main researcher presenting the data, Dr. Thomas Verstraeten. After giving a brief background of his work, Verstraeten launched into his findings:

> For the overall category of neurologic developmental disorders, the point estimates of the categorized estimates suggest potential trends, and the test for trends is also statistically significant above one, with a P value below 0.01. The way to interpret this point estimates which seems very low is as follows: That's an increase of .7% for each additional microgram of ethylmercury. For example, if we would go from zero to 50 micrograms of ethylmercury, that would give us an additional increase of about 35%, which is pretty close to the point estimate for this category. Or, for the

overall, we would have to multiply 75 micrograms to .7 and that would give us about a one and a half to the relative risk.[6]

This meant that for each additional microgram of mercury added with a vaccine, there would be a .7 percent increase in the number of children with a neurological disorder. Given that it was estimated that those who were most aggressively vaccinated would receive around 75 micrograms of mercury, the subsequent risk for a neurological disorder went up 50 percent.

There was some discussion of the varying factors which the researchers might want to consider, such as low birth weight and mercury exposure of the mother during pregnancy, but eventually Verstraeten felt it necessary to summarize his findings:

> The bottom line to me is you can look at this data and turn it around and look at this and add the stratum. I can come up with risks very high. I can come up with risks very low, depending on how you turn everything around. You can make it go away for some and then it comes back for others. If you make it go away here, it will pop up again there. So the bottom line is, okay, our signal will simply not go away.[7]

The signal simply would "not go away" no matter how much they tried to change the parameters of the study. The only logical conclusion is that they were dealing with something real, and it was affecting the neurological development of children. Verstraeten continued:

> Personally, I have three hypotheses. My first hypothesis is it is parental bias. The children that are more likely to be vaccinated are more likely to be picked up and diagnosed.
>
> Second hypothesis, I don't know. There's a bias that I have not yet recognized, and nobody's told me about it.

Third hypothesis, it's true, it's thimerosal. Those are my
hypotheses.[8]

Verstraeten was then challenged by another researcher as to the possible mechanisms and biological plausibility of injected mercury causing
neurodevelopmental problems in children, and he replied:

When I saw this and I went back through the literature, I was
actually stunned by what I saw because I thought it is plausible.
First of all, there is the Faroe study, which I think people have
dismissed too easily. And there is a new article in the same journal that was presented here, the *Journal of Pediatrics*, where they
have looked at PCB. They have looked at other contaminants in
seafood and they have adjusted for that, and still mercury comes
out. That is one point.

Another point is that in many of the studies with animals, it
turns out there is quite a different result depending on the dose of
mercury. Depending on the route of exposure and depending on
the age at which the animals were exposed. Now, I don't know how
much you can extrapolate that from animals to humans, but that
tells me that mercury at one month of age is not the same as mercury at three months, at twelve months, prenatal mercury, later mercury. There is a whole range of plausible outcomes from mercury.[9]

It had been a remarkable first day and the invited scientists and medical
personnel had been given a great deal to think about.

It was on the second and final day of the conference that the real
fireworks began.

* * *

The second day started with many experts pouring cold water on
the findings of Dr. Verstraeten, although without the benefit of any

significant data. To characterize their responses, it seemed they "just didn't believe the results."

The justifications would come later.

Typical of these challenges were the remarks of Dr. Paul Stehr-Green, an associate professor of epidemiology at the University of Washington School of Public Health and Community Medicine:

> First, I want to reiterate what others have said. I want to congratulate the folks who did the initial analyses for a tremendous amount of work, a lot of dedication, and very interesting results. In my judgment, these preliminary results are not compelling. But the implications are so profound that the lead should be examined further. My outstanding concerns and reasons for that statement really go to the validity and accuracy of these results that revolve primarily around the issue of ascertainment bias or confounding, which I think is potentially a fatal flaw, which was not dispelled by some of the clever analyses.[10]

It's fortunate I'm trained in the language of science from my years as a middle school science teacher and can translate for the casual reader. He really meant to say, "Wow, you did lots of work, and even though you looked at the information in many different and clever ways, I still don't believe you."

His point was partially echoed by Dr. David Johnson, the public health officer for the state of Michigan, but he went further and actually suggested something reasonable and responsible:

> In my opinion, the evidence today is insufficient to determine whether or not thimerosal containing vaccines caused the neurological sequalae in question. The diagnosis, even in the hands of experts, and the number of diagnoses are too easily influenced by variations in parental and physician sensitivity and concern, and utilization of health care of similar merits. The underlying

biologic, toxicologic and pharmacologic data are too weak to offer guidance one way or the other.

Now, on the other hand, the data suggests that there is an association between mercury and the endpoints, ADHD, a well-known disability and speech delay, as entered into the database. Then, here comes an opinion. Well, it is all an opinion, but it expresses a flavor, so I think it is related to what Dr. Bernier [associate director for the National Immunization Program] is trying to derive here.

This association leads me to favor a recommendation that infants up to two years old not be immunized with thimerosal containing vaccines if suitable alternative prescriptions are available. [Bold and italics by author.][11]

However, nobody in the public was ever allowed to consider this reasonable suggestion. The reason is probably because if the public health establishment made such a recommendation, many in the public would reasonably ask, what makes you think it's safe after age two?

Maybe it should be age three, or four, or five.

Or if the public wanted to propose a truly challenging scientific question, they might ask, is it ever a good idea to have mercury in a product which you inject into a person's bloodstream?

But the media never let the good people of the year 2000 have those conversations. How different our world might look today if they had.

However, it seemed most of the participants at the Simpsonwood Conference weren't worried about the harm to children. They were worried about lawyers and lawsuits. These are the comments of Dr. Robert Brent, developmental biologist and pediatrician from Thomas Jefferson University and the Dupont Hospital for Children:

The medical/legal findings in this study, causal or not, are horrendous and therefore it is important that the suggested epidemiological, pharmacokinetic, and animal studies be performed. If an allegation was made that a child's neurobehavioral findings were caused by thimerosal containing vaccines, you could readily find a junk scientist who would support the claim with "a reasonable degree of certainty." But you would not find a scientist with any integrity who would say the reverse with the data that is available. And that is true. So, we are in a bad position from the standpoint of defending any lawsuits if they were initiated, and I am concerned.[12]

Maybe I'm missing something, but I would have thought a developmental biologist and pediatrician from Thomas Jefferson University and the Dupont Hospital for Children would be more concerned about children than lawsuits aimed at vaccine makers, or the public agencies which recommended them for children.

In other words, nobody was going to sue Dr. Robert Brent.

They'd go after the pharmaceutical companies, or failing that, they'd go after the CDC and the National Institutes of Health.

It kind of makes you wonder to whom Dr. Brent, and others like him, owed their loyalty.

Much of the discussion continued in this vein, with a small number saying it was time to inform the public, while the majority wanted to keep the information quiet until more research could be conducted.

But then came the concluding speech.

* * *

The most distinguished member of the group assembled at the Simpsonwood Resort was, without a doubt, Dr. John Clements. At the time, he was seventy-seven years old and working for the Expanded Program on Immunization for the World Health Organization. In 1994, he had won the Lasker Award, often referred to as the American Nobel Prize for medical research. In 2008, he would win the Pollin

Prize for Pediatric Research, the only international pediatric award for research.

Dr. Clements was asked by the chairman to give his comments on "the implications of dealing with the composition of vaccines for the international community." Clements thanked the chairman and then stood to speak:

> First of all, I want to thank the organizers for allowing me to sit quietly at the back. It has been a great privilege to listen to the debate and to hear everybody work through with enormous detail. And I want to congratulate, as others have done, the work that has been done by the team. Then comes the BUT.
>
> I am really concerned that we have taken off like a boat going down one arm of the mangrove swamp at high speed, when in fact there was not enough discussion early on about which way the boat should go at all. And I really want to risk offending everyone in the room by saying that ***perhaps this study should not have been done at all***. [Bold and italics added by author.]
>
> Because the outcome of it could have, to some extent, been predicted. And we have all reached this point now where we are leg hanging [over the side of the boat?], even though I hear the majority of the consultants say to the board they are not convinced there is a causality direct link between thimerosal and various neurological outcomes.
>
> I know how we handle it from here is extremely problematic. The ACIP [Advisory Committee on Immunization Practices] is going to depend on comments from this group in order to move forward into policy. And I have been advised that whatever I say should not move into the policy area because that is not the point of this meeting.
>
> But nonetheless, we know from many experiences in history that the pure scientist has done research because of pure science. But that pure science has resulted in the splitting of the atom or

some other process, which is completely beyond the power of the scientists who did the research to control it. And what we have here is people who have, for every best reason in the world, pursued a direction of research.

But there is now the point at which the research results have to be handled, and even if this committee decides that there is no association and that information gets out, the work has been done. And through Freedom of Information [Act] that will be taken by others and will be used in other ways beyond the control of this group. And I am very concerned about that as I suspect it is already too late to do anything, regardless of any professional body and what they may say.

My mandate as I sit here in this group is to make sure that at the end of the day that one hundred million are immunized with DTP, hepatitis B, and if possible HiB, this year, next year, and for many years to come. And that will have to be with thimerosal-containing vaccines unless a miracle occurs and an alternative is found quickly and tried and found to be safe. [Bold and italics added by author.]

So, I leave you with the challenge that I am very concerned that this has gotten so far, and having got this far, how you present in a concerted voice the information to the ACIP in a way that they will be able to handle it and not get exposed to other traps which are out there in public relations. My message would be that any other study, and I like the study that has been described very much, I think it makes a lot of sense, but it has to be thought through.

What are the potential outcomes and how will you handle it? How will it be presented to a public and media that is hungry for selecting the information they want to use for whatever means they have in store for them?

I thank you for that moment to speak, Mr. Chairman, and I am sorry if I have offended you. I have the deepest respect for the work that has been done and the deepest respect for the analysis that has been done. But I wonder, how on Earth you are going to handle it from here.[13]

And with that question, Dr. Clements, working for the Expanded Program on Immunization of the World Health Organization, recipient of the Lasker Award, and future winner of the Pollin Prize for Pediatric Research, took his seat.

As I read the words of Dr. Clements, it seemed he was engaging in the vaccine two-step. On one hand, he compliments the researchers for their work and analysis, then reverses course by questioning whether the research should have been done at all because "the outcome of it could have, to some extent, been predicted."

Has there ever been a comparable moment in science?

Can you imagine a gathering of Roman Catholic bishops in the seventeenth century saying to Galileo, "Really first-class observations and mathematics about how the Earth revolves around the Sun, but have you considered how it will play among the public and the media? We know we can't disprove your theory because the facts seem to be on your side, but this definitely needs to be handled."

Even the entitled princes of the Catholic Church in the Middle Ages did not go as far as the scientists who gathered at the Simpsonwood Retreat Center during those two days in June of 2000, when they decided to cover up potential harm to children.

* * *

I will confess my bias.

I lay the battles of the past two decades about vaccines directly at the feet of the fifty-three individuals who met for those two days in June 2000 at the Simpsonwood Retreat Center in Norcross, Georgia.

It is clear to me that prior to that meeting, leading scientists did not believe there was anything wrong with the vaccination schedule and would have had contempt for anybody who made such an assertion. After that meeting, there was no doubt that something had gone wrong with childhood vaccines.

I am not privy to the inner thoughts of leading scientists. But from the actions which followed I can only conclude that a decision was made to never allow an open public discussion of these facts and issues, the way the scientists at Simpsonwood were allowed to speak.

Simpsonwood is the place where the contempt among scientists about vaccine injury shifted to an active conspiracy to cover up the information and viciously attack anybody who was genuinely trying to protect the public.

The public health establishment of the United States did not act in the best interest of its citizens.

Our son, Ben, was born approximately two weeks after the Simpsonwood conference, on June 26, 2000. He was on the regular vaccine schedule up until January 2002, when his eighteen-month series of shots left him mute for twelve days. We never gave him another shot.

Our daughter, Jacqueline, who was born in 1998, had received the recommended shots for a three-and-a-half year-old. By that time, she had a diagnosed seizure disorder and autism. She was born healthy, regressing only after her six-month series of shots, which we were only able to piece together when we saw the dramatic change in our son.

Our son is now twenty-one years old, looking to graduate from college in 2022, has a girlfriend, and a job in accounting waiting for him.

Our daughter, twenty-three, still cannot speak, and needs help with most of her daily living activities. Unless modern medicine steps up to the challenge, she will probably be living with my wife and me until the day we die.

Not one of the fifty-three members of the medical establishment at the Simpsonwood Conference broke ranks and shared their story with the *New York Times* or *Washington Post*.

Because of their failure to act, we have now lost more than twenty-one years of medical and scientific research which might have been able to fix many of these problems.

Sadly, the conspiracy continues to this day.

CHAPTER SIX

Whistleblowers Used to Stop Wrongdoing—Does Anybody Care About Them Today?

There once was a time in American history when an allegation of wrongdoing by the government would dominate the headlines. Consider the 1971 Pentagon Papers case, which revealed decades of government lies about the war in Vietnam, from the final days of World War II until May 1968, all courtesy of Daniel Ellsberg, a senior research associate at the Massachusetts Institute of Technology's Center for International Studies and their publication by the *Washington Post*.

As described by the Encyclopedia Britannica:

> The Pentagon Papers revealed that the Harry S. Truman Association gave military aid to France in its colonial war against the communist-led Viet-Minh, thus directly involving the United States in Vietnam; that in 1954 President Dwight D. Eisenhower decided to prevent a communistic takeover of South Vietnam and to undermine confidence in the new communist regime of North Vietnam; that President John F. Kennedy transformed the policy "limited-risk gamble" that he had inherited into a policy

of "broad containment"; that President Lyndon B. Johnson intensified covert warfare against North Vietnam and began planning to wage overt warfare in 1964, a full year before the depth of U.S. involvement was publicly revealed; and that Johnson ordered the bombing of North Vietnam in 1965 despite the judgment of the U.S. intelligence community that it would not cause the North Vietnamese to cease their support of the Viet Cong insurgency in South Vietnam.[1]

As our country becomes more and more politically divided, it becomes difficult to realize the members of our political team may be just as corrupt as we accuse the other side of being. Truman lied about Vietnam. Eisenhower lied about Vietnam. Kennedy lied about Vietnam. And Johnson lied about Vietnam. That's more than two decades of lies told by American presidents, Republican and Democrat. Does there really appear to be much difference between the parties, despite what their rhetoric might claim? What are we to make of this? Is there only one true party, controlling both sides of our politics? It's difficult to come to any other conclusion.

However, it used to be that a truth-teller had at least the potential to destabilize the edifice of lies erected by both political parties. And there were enough people willing to listen to the possibility that their own favored candidate was just as guilty as the other side:

> The disclosures and their continued publication [by the *Washington Post*] despite top-secret classification were embarrassing to the administration of president Richard M. Nixon, who was preparing to seek reelection in 1972. So distressing were these revelations that Nixon authorized unlawful efforts to discredit Ellsberg, including burglarizing the office of Ellsberg's psychiatrist in an attempt to unearth embarrassing information. Those efforts came to light during the investigation of the Watergate scandal.[2]

The Supreme Court sided with the *Washington Post* in a 6–3 decision and even the president of the United States could not stop their publication. However, the public loss of face by Nixon caused him to engage in activities which ultimately led to his resignation in August 1974.

That was the power of a whistleblower in the 1970s.

As I write these words in mid-August 2021, it has been just a few days since the fall of Afghanistan to the Taliban. Since the invasion of Afghanistan in October 2001, we have been told that we are beating the Taliban. More recently, we have been consistently reassured that when we left, the more than three thousand members of the Afghan Security Forces that we spent twenty years training would be able to prevent the Taliban from returning to power.

In that time, we have heard the same optimistic talk from President George W. Bush, President Barack Obama, President Donald Trump, and President Joe Biden.

It seems beyond dispute that our leaders consistently lie to us about the success of our military efforts abroad.

If they lie about important foreign issues, is it possible they lie about domestic issues, like public health, as well?

* * *

It all started with Dr. William Thompson, a senior scientist at the Immunization Safety Division of the CDC, calling the wrong number on November 7, 2013.

According to Thompson, he thought he was calling the office of United States Senator Patty Murray (D) Washington. But instead of Senator Patty Murray, Thompson was calling Dr. Brian Hooker, a fifty-year-old associate professor of biology and chair of the Math and Science Division of Simpson University in Redding, California.

Thompson and Hooker had talked a great deal between 2002 and 2003, when Dr. Hooker was part of more than five thousand families who had sued the US government over the development of autism in their children, which they attributed to vaccines, in what was known as the Autism Omnibus Group. Because of his scientific background, Hooker had become the de facto spokesperson for the parents, while Thompson acted as the CDC point of contact.

However, in 2003, Hooker had been told that since he was now an "adversary" of the CDC, his communications with Thompson would need to end. When Hooker found that Thompson had called after nearly a decade of silence, he was in no mood for polite conversation. An exchange of emails followed, with Thompson saying the call had been a mistake. Hooker emailed back:

> Bill,
>
> Your account of the call makes no sense. A seasoned government scientist like yourself would know that DC numbers for Congress start with a 202 code (224 for the Senate and 225 prefix for the House). Also, if you want to call Senator Murray's office, why wouldn't you simply look up her number at murray.senate.gov?
>
> Could you please tell me the "real reason" you were trying to get in touch with me by phone? I don't have time for more CDC lies.
>
> Brian.[3]

More emails ensued between the two men and it seemed to Hooker as if Thompson had become something of a different man. Thompson agreed with Hooker about several scientific papers that disputed a link between vaccines and autism. Thompson agreed the papers were not as strong as the authors and scientific media claimed.

In their first phone call on November 13, 2013, Thompson mysteriously told Hooker that he wanted to talk to him, but "not while he was on CDC property."[4]

That conversation took place a few days later, and I wrote about it in my 2020 book, *Inoculated: How Science Lost Its Soul in Autism*, which was essentially co-written with Dr. Brian Hooker. [I offered Hooker a co-writing credit because of his enormous help in not only providing information, but also editing and fact-checking my work. He declined. Hooker did not want to profit in any way from the revelations of Dr. Thompson.]

> On Friday, November 15, 2013, Bill Thompson called Brian Hooker for their second conversation. After a few minutes of social chitchat, Thompson got down to the purpose of his call: "You're going at it all wrong with the Geiers [fellow collaborators] and trying to get into the Vaccine Safety Data-Link. Why are you doing that? That's just the wrong way to go. You need to be requesting the public use dataset. They're publicly available and they're available to you by law. You need to go through a particular procedure, and I can give you the email of the individual to contact. They have to give you these data-sets if you request them. But you have to do it the right way."
>
> Hooker was stunned by the sudden revelation and the urgency in Thompson's voice. "Okay," he replied, "I had no idea you could do that."
>
> Something must have broken in William Thompson, because in that moment it seemed to Hooker that this CDC employee had decided to change from being a bureaucrat to operating as a scientist whose sole obligation was to tell the truth.
>
> "If you follow my lead," I will guide you through this," said Thompson. "You will have more data than you know what to do with."[5]

Hooker did as Thompson instructed him, but searched first for evidence of mercury and the development of autism. Thompson suggested he look instead for the dataset from the Frank DeStefano study, published in 2004, which had been the death-knell to most in the scientific community regarding the claims of Dr. Andrew Wakefield that the MMR (measles-mumps-rubella) vaccine was linked to the development of gastrointestinal problems and autism.

Hooker requested that dataset, it was provided to him, and he started to work on it:

> The first data analysis Hooker ran was on children who received the MMR vaccine before thirty-six months and those who received the MMR vaccine after thirty-six months. The odds ratio for the earlier group was a 1.49 increase over those who received the MMR shot after thirty-six months. The CDC had actually published that number in the DeStefano study. This was a statistically significant number. But the CDC explained it away by saying that children diagnosed with autism were receiving special services that required them to get an MMR shot. Hooker knew this explanation was preposterous. In the early 2000s, an autism diagnosis was rarely made before a child was three years old.
>
> Hooker then ran the analysis on just boys, as the rate of autism is known to be higher in males than females, and found the odds ratio went up to 1.67. The CDC had also reported this number in the DeStefano study. When he ran the analysis on the girls, he was not able to find an increased odds ratio. Hooker wryly noted that the DeStefano study authors had not commented on how their earlier explanation for the increase due to inclusion in special education services failed to explain the negative finding in girls.
>
> The next analysis Hooker utilized was for African American boys and girls. The odds ratio was a 2.6-fold increase. In science, an odds ratio above 2.0 is considered a clean and convincing signal and is often considered proof of causation in a legal case.
>
> Hooker then analyzed the effect for the African American boys and found the odds ratio jumped to 3.36.
>
> This was a smoking gun.
>
> He went back to the DeStefano paper and noticed they had done something unusual with the African American group members who had a valid State of Georgia birth certificate. As Hooker

calculated it, the CDC scientists had thrown out about ninety of the 220 African American children with autism, lowering the pool by 40 percent and dramatically skewing the odds ratio.

Hooker compiled the information, checked and double-checked his numbers, typed up the results, and then scheduled a conference call with Thompson.

When Hooker got Thompson on the phone, he told him what his analysis had revealed . . .

"How long did it take you to find it?" Thompson asked.

"About thirty minutes after I started programming."

"Yeah, it just jumps off the page, doesn't it?"[6]

While it's devastating to think of any child developing autism, it's mind-boggling to consider the harm this single shot was causing to the Black male population. In just about any court in this country, if there was an odds ratio of 3.36, meaning a more than three-fold increase, judgment would be swift and severe.

Dr. Hooker was so alarmed by the data he felt he had to reach out to Dr. Andrew Wakefield, whose work had first raised concerns about the MMR shot, but who had then been subjected to a campaign of unprecedented vilification.

Wakefield, who'd had his own experiences with whistleblowers losing their courage, told Hooker he needed to lock Thompson's story in. They consulted an attorney who recommended Hooker travel to Washington State and call Thompson on the phone and get this information recorded. Washington allowed one party consent to the taping of a phone call. Hooker did this and recorded several conversations.

Probably one of the most damaging conversations came from a recording on May 24, 2013, in which Thompson said:

> You see that the strongest association is with those [autistic] cases without mental retardation. The non-isolated, the non-MT [mental retardation] . . . the effect is where you would think it would

happen. It is with the kids without other conditions, without the comorbid conditions . . . I'm just looking at this and it's like, "Oh my God!" . . . I cannot believe what we did. But we did. . . . It's all there . . . It's all there. I have handwritten notes.[7]

As it turns out, at this time, Thompson seemed to be quite open to testifying about the actions he'd engaged in more than a decade earlier. In fact, on August 27, 2014, Thompson released a letter to the public through his lawyers, about the allegations. It seemed to be a complete confession. I have reproduced the majority of his statement.

> **For Immediate Release—August 27, 2014**
> **Statement of Willim M. Thompson, Ph.D., Regarding the 2004 Article Examining the Possibility of a Relationship Between MMR Vaccine and Autism**
> My name is William Thompson. I am a senior scientist with the Center for Disease Control and Prevention, where I have worked since 1998.
>
> I regret that my coauthors and I omitted statistically significant information in our 2004 article published in the journal *Pediatrics*. The omitted data suggested that African American males who received the MMR vaccine before age 36 months were at increased risk for autism. Decisions were made regarding which findings to report after data was collected, and I believe the final study protocol was not followed.
>
> I want to be absolutely clear that I believe vaccines have saved and continue to save countless lives. I would never suggest that any parent avoid vaccinating children of any race. Vaccines prevent serious diseases, and the risks associated with their administration are vastly outweighed by their individual and societal benefits.
>
> My concern has been the decision to omit relevant findings in a particular study for a particular subgroup for a particular

vaccine. There have always been recognized risks for vaccination and I believe it is the responsibility of the CDC to properly convey the risks associated with the receipt of those vaccines.

I have had many discussions with Dr. Brian Hooker over the last 10 months regarding studies the CDC has carried out regarding vaccines and neurodevelopmental outcomes including autism spectrum disorders. I share his belief that CDC decision-making and analysis should be transparent. I was not, however, aware that he was recording any of our conversations, nor was I given any choice regarding whether my name would be made public or my voice would be put on the internet.

I am grateful for the many supportive emails that I have received over the last several days. I will not be answering further questions at this time. I am providing information to Congressman William Posey, and of course will continue to cooperate with Congress.[8]

In just about any other situation in the world a confession such as this would mean the case was open and shut. He confessed to omitting "statistically significant information" about harm to African American boys. He admits he was in regular contact with Dr. Hooker and that it was his voice that was recorded. Where was the *New York Times* or *Washington Post*, ready to swoop in and investigate this case of medical racism?

Maybe there was no rush because Big Pharma buys itself protection with all of its marketing dollars.

The claims of Dr. Thompson about harm to African American boys are not a theory, but admitted fact.

After examining the documents, Congressman William Posey felt he had to speak about what he had discovered, which he did on July 29, 2015, including a letter written by Dr. Thompson. This is Congressman Posey on the floor of the House of Representatives, quoting from Dr. Thompson's letter:

My primary job duties while working in the Immunization Safety Branch from 2000 to 2003 were to lead or co-lead three major vaccine safety studies. The MADDSP-MMR Autism Cases-Control study was being carried out in response to the Wakefield Lancet Study that suggested an association between the MMR and an autism-like health outcome. There were several major concerns among scientists and consumer advocates outside the CDC in the fall of 2000 regarding the execution of the Verstraeten study. One of the important goals that was determined up from in the spring of 2001 before any of these studies started, was to have all three protocols vetted outside the CDC prior to the start of analyses, so that consumer advocates could not claim we were presenting analyses that suited our own goals and biases. We hypothesized that if we found statistically significant effects, at either 18- or 36-month thresholds, we would consider that vaccinating children early with MMR vaccine could lead to autism-like characteristics or features.

We all met and finalized the study protocol and analysis plan. The goal was not to deviate from the analysis plan to avoid the debacle that occurred with the Verstraeten Thimerosal Study published in *Pediatrics* in 2003. At the September 5th meeting we discussed in detail how to code race for both a sample and the birth certificate sample. At the bottom of Table 7 it also shows that for the non-birth certificate sample, the adjusted-race effect, statistical significance was HUGE.

All the authors and I met and decided sometime between August and September 2002 not to report any race effects for the paper. ***Sometime soon after the meeting we decided to exclude reporting any race effects, the coauthors scheduled a meeting to destroy documents related to the study. The remaining four coauthors all met and brought a big garbage can into the meeting room and reviewed and went through all the hard copy documents that we had thought we should discard and put them in a big garbage can.*** [Bold and italics added by author.]

> However, because I assumed it was illegal and would violate
> both FOIA [Freedom of Information Act] and DOJ [Department
> of Justice] requests, I kept hard copies of all documents in my
> office, and I retained all associated computer files. I believe we
> intentionally withheld controversial findings from the final draft
> of the *Pediatrics* paper.[9]

It seems incomprehensible that scientists could act with such illegality
in the face of evidence that a medical intervention was having such a
devastating effect on the Black population. Were these scientists "white
supremacists," as Justice Oliver Wendell Holmes Jr. was when he wrote
the majority opinion in *Buck v. Bell*?

Or was there something even more nefarious at work? A disdain for
the human race, perhaps not simply on the basis of race, but creating
one class of people who know the truth, and the other who are told
lies? I'm not even sure we have a name for such a concept in the English
language. I'm tempted to call them elitists.

And yet the word sounds completely inadequate to describe the
situation.

I'd like to tell you this story has a happy ending—that we got to
the bottom of why African American boys developed autism at higher
rates when given the MMR shot before the age of thirty-six months,
and that a complete investigation was done of the vaccine schedule.
Given the fact we're talking about the health of children, maybe the
CDC even did a study looking at the health outcomes of vaccinated
children and non-vaccinated children.

But no such study has ever been done by the CDC.

And no charges have ever been filed against Dr. William Thompson,
or any of his coauthors who got together at the CDC on a Saturday
afternoon in 2002 to destroy documents.

* * *

The removal of vaccines from the regular civil justice system was accompanied by the promise of a quicker mechanism to get money to the families of vaccine injury victims, while at the same time ensuring that the public got accurate information about the risks of vaccines.

Dr. Brian Hooker was among the five thousand families who sought relief under the Vaccine Court in what became essentially a class action under what was called the Omnibus Autism Proceeding. As previously mentioned, these claims were denied in 2008.

But was justice done in these cases?

In January 2019, reporter Sharyl Attkisson broke a story in *The Hill*, which many believed would finally bring an end to decades of lies about the vaccine-autism connection. The article was titled, "How a Pro-Vaccine Doctor Reopened Debate about Link to Autism."[10]

According to Attkisson, the government's own medical witness, Dr. Andrew Zimmerman of Johns Hopkins University, was ready to testify in 2007 that at least a third of the autism cases were the result of vaccine injury, but he was prevented from testifying. From Attkisson's article:

> Dr. Zimmerman has now signed a bombshell sworn affidavit. He says that, during a group of 5,000 vaccine-autism cases being heard in court on June 15, 2007, he took aside the Department of Justice (DOJ) lawyers he worked for defending vaccines and told them he'd discovered "exceptions in which vaccines could cause autism."
>
> "I explained that in a subset of children, vaccine-induced fever and immune stimulation did cause regressive brain disease with features of autism spectrum disorder," Dr. Zimmerman now states. He said his opinion was based on "scientific advances" as well as his own experience with patients.
>
> For the government and vaccine industry's own pro-vaccine expert to have this scientific opinion stood to change everything about the vaccine-autism debate, if people were to find out.

But they didn't.

Dr. Zimmerman goes onto say that once the DOJ lawyers learned of his position, they quickly fired him as an expert witness and kept his opinion secret from other parents and the rest of the public.[11]

Sometimes the litany of lies told about vaccines can be so overwhelming it causes people to turn away because it is difficult to believe scientists can be so corrupt. I understand this feeling because as I have researched this topic over the past eighteen years there have been many times I wanted to turn away from the horror. The Hippocratic Oath begins with the words, "First, do no harm."

However, we must be clear-eyed about the problems.

I have shown how, in 2000, the CDC detected a signal from their vaccine surveillance program and convened a special meeting of top researchers in Simpsonwood, Georgia and made the decision to bury the information.

I have detailed how, in 2002, senior scientists from the CDC made the decision to destroy information showing an increased risk of autism in Black boys from earlier administration of the MMR vaccine, a single act which has likely destroyed the lives of tens, if not hundreds of thousands.

I have given information about how, in 2007, the government's own medical witness in the Omnibus Autism Proceeding was prevented from testifying that vaccines can and do cause autism.

How can we have *any* faith in our political, legal, and pharmaceutical leaders in light of this information?

CHAPTER SEVEN

The Chuck Todd Affair— How the Media Acts as an Assassin for Big Pharma

My second book with Dr. Judy Mikovits, *Plague of Corruption*, was published on April 14, 2020.

On May 10, 2020, less than a month after publication, it hit number one in all books sold on the Amazon charts in non-fiction, beating out Michelle Obama's book, *Becoming*.[1] That's in addition to later being on the *New York Times* bestseller list for six weeks (eventually reaching #2 for the week of June 14, 2020) and a bestseller for the *Wall Street Journal*, *Publishers Weekly*, and *USA Today*. And to make it even sweeter, our 2014 book *Plague* was number 10 on the Amazon charts for most books sold in a week.

We were truly cornering the market on explaining viruses to the public, something they were interested in with the COVID-19 crisis and the nationwide lockdown starting in March. Many called our book "prophetic," but to me they were just seeing the same patterns of corruption. The same people who lied about Judy's work were lying about COVID, its danger, where it came from, and how to effectively combat it.

By the end of 2020, we had sold more than two hundred thousand copies, making it one of the most popular science books of the year.

One might have thought we had a massive marketing program, glossy displays and flattering profiles in all the top newspapers and magazines, and perhaps even fawning interviews on *Good Morning America* or *Fox & Friends*.

But we didn't have any of that.

I had come up with a scrappy, low cost marketing plan using some of the biggest voices in conservative media, such as cartoonist Ben Garrison. I commissioned a cartoon from Garrison to appear on the day of our book's release, entitled "Real Science Defenders." Please understand my coauthor Judy was trained as a scientist, weighing and evaluating data, and trying to make logical connections supported by the evidence. Cartoons to get a point across? She didn't get it at first.

But it was clear to me the sober, scientific world she thought existed was instead controlled by large corporations, intent on squashing whis-tleblowers like her.

I often told her she needed to think differently, like a southpaw being trained to punch more effectively with his right.

This was guerilla warfare with the health of humanity hanging in the balance.

The cartoon by Ben Garrison, which he tweeted out to his more than two hundred thousand followers, was done in the style of an Avengers movie poster. At the center was a large, green, two-legged creature with the body of a weight-lifter, feet with menacing claws, and a lizard-like tail with deadly vaccine syringes at the end like some armor-plated dinosaur or monster of legend. The creature wore red shorts and a belt buckle identifying it as "Big Pharma."

The head of the monster was a prescription pill bottle, a jack o'lan-tern pumpkin face, and the letters NIH (National Institutes of Health) on the cap. Judy was depicted as a flying superhero with a red cape, yellow boots, and gloves to match my flowing blonde hair, all of which was packaged in a blue body suit.

It reminded Judy of the birthday card one of her patients made, depicting her as Wonder Woman, with the caption, "If I don't get to wear pants, nobody gets to wear pants."

To this day, her husband David laughs and says, "Beware of commando Judy!"

With one curled fist Judy's punching the great beast, the prescription cap popping off to reveal a dazed Dr. Francis Collins (head of the NIH), and in her other fist she holds a copy of our book. Flying above her in support are her three fellow superheroes: me, breaking a large vaccine syringe, Robert Kennedy Jr., holding up a torch which reads "Truth," and Dr. Andrew Wakefield, holding a shield into which five syringes have been stopped, and pointing at the creature with the other hand and shouting, "Stop!"

In one hand the creature holds a cage with a sad looking mouse in it (a nod to Judy's discovery of the mouse retrovirus, XMRV, in patients with chronic fatigue syndrome [ME-CFS]), and in the other he holds a giant syringe in which floats a small mouse and aborted baby (a nod to my question as to whether it was ever a good idea to inject any biological material into the bloodstream of a human being.)

In the left corner of the picture is an open sewer grate and crawling out of it is a large green worm with the head of California Governor Gavin Newsom, and a syringe, saying, "It's the law," in reference to the mandatory vaccine law he'd signed, eliminating religious and philosophical objections.

In the bottom right corner, carrying a bag of money and trying to run away, is Dr. Anthony Fauci, the true Dr. Frankenstein in this horror story.

After the book came out, many claimed we were taking shots at Dr. Fauci because of his recent celebrity position as the head of Trump's coronavirus task force. But that misses the fact that we'd heavily criticized him in our 2014 book *Plague*, depicting him as the puppet master making XMRV go away. And as for featuring him in our April 2020 release, the book was finished in November 2019 before most people knew the name Anthony Fauci.

Either we were incredibly lucky, or we knew what we were talking about.

But many others had made the connection, which wasn't difficult since Fauci had been head of the Institute of Allergy and Infectious Diseases (NIAID) since 1984. That meant he'd served as head of his agency longer than the corrupt J. Edgar Hoover had served as head of the Federal Bureau of Investigations (FBI). During a lunch with Kris Newby, the Stanford Medical School science writer, author of the book *Bitten*, and director of the documentary *Under Our Skin*, both detailing the history of Lyme disease, she'd told me that Fauci's long reign was the reason so many chronic conditions hadn't been properly investigated. It was Newby's opinion that Fauci, not Francis Collins, the head of the NIH, was the true emperor of public health, controlling everything because of his role in approving grant applications.

After Ben Garrison's cartoon, which I'd been planning for several months, came Darla Shine, the wife of Bill Shine, the former president of Fox News for nineteen years, and for a time the White House communications director for President Trump. Darla was known as an important social media influencer and author, and she had a personal connection to the vaccine issue. "I don't often tell people this," she said to me, in an early conversation, "but here I was married to the president of Fox News, and up until 2016 I was a Democrat." She posted a picture of our book next to a big glass of red wine, telling people she was planning to sit down and read it. That helped a great deal.

Author, Michelle Malkin, who has a vaccine-injured child, was also an early and enthusiastic backer. I contacted her and sent her a book. On April 16, 2020, she sent out a tweet of herself holding the book and wrote in the accompanying text, "Anthony Fauci, Bill Gates and #BigPharma don't want you to read this book. Get #PlagueOfCorruption now and learn truth about public health industrial complex's decades of research fraud and vaccine cover-ups."[2] The tweet went out to her 2.2 million followers, and was hugely popular, garnering, as of December 2020, 213 comments, 3,141 retweets, and 4,858 hearts. In addition, Michelle was gracious enough to invite us onto her podcast, where we were able to discuss the book at length.

Michelle quickly understood our view, that despite us having some support from conservatives, most of the conservative and liberal news media were opposed to us because of Big Pharma money.

We did an appearance on the nationwide Coast to Coast AM radio show, hosted by George Noory, on April 13, 2020, the Monday night before our Tuesday release. We'd appeared on Coast to Coast a few times before and George was always exceedingly kind.

But would it be enough to alert the public to our dire message?

* * *

Besides commissioning a cartoon from Ben Garrison, I thought it would be worth the money to hire a publicist. I got a recommendation from another author, Marcia Hinds, to use Michael Wright of Garson and Wright Public Relations Services and Literary Management.[3] They weren't cheap, but were supposed to be the best in the business, particularly with controversial books.

They'd handled high profile books by Geraldo Rivera, O. J. Simpson's *If I Did It*, a book about the accused murdering mom Casey Anthony, an account of the Columbine High School shooting, and a book by Rachel Dolezal, president of the local chapter of the National Association for the Advancement of Colored People in Spokane, Washington, who lost her job when it was revealed she was not Black, but only self-identified as Black.

All these books got national media exposure for their authors.

"Michael, I only give you about a 30 percent chance of being able to get us national media," I told him early in the relationship. "I recognize I'm probably throwing this money down the toilet."

Michael was more confident. "It's not going to be any problem getting you on Tucker Carlson," he said, "I just want to see if we can get you onto some of the more mainstream news shows like NBC, before we go to Fox."

"You're the expert," I replied.

Michael also promised a "National Radio Tour," which didn't mean touring the country, but instead meant spending eight hours in a radio room in some urban area, and doing ten to fifteen-minute interviews, on a call-in line. By the end of the day we would have hit many of the biggest cities in the country.

We never got on NBC News.

We never got on Tucker Carlson.

And we never did get to do that National Radio Tour.

It doesn't seem like there's a high demand for the truth in today's media.

* * *

In the Ben Garrison cartoon, he'd depicted Big Pharma as an enormous beast. But what were the characteristics of this beast, and how large, exactly, was it? It can be difficult to find good information. But there have been some informative publications over the years. Consider this article from CBS News in March 2016:

> The biggest game at this year's Super Bowl didn't take place on the field. It was played out during the commercial breaks when three pharmaceutical companies took to the air.
>
> But those costly TV spots—for issues including irritable bowel syndrome and toenail fungus—represented just a fraction of the $5.2 billion in annual advertising spending from the drug industry . . .
>
> . . . In November, the American Medical Association took the unusual step of calling for a ban on pharmaceutical ads. The reason? Doctors believe the surge in drug ads is prompting consumers to demand expensive medications they might not need.[4]

We begin to get a picture of the size of the beast. Three pharmaceutical companies could use the Superbowl, our most popular national event, to advertise their products. Doesn't that make them among our largest companies? However, with their peaceful advertisements for the better life we'll get with their drugs (but don't mind the side-effects, which might include death) everything will be simply fine and dandy.

And if that doesn't bother you, well, apparently, it concerned the American Medical Association enough to call for a ban on the ads. That's the organization to which practically every single medical doctor in the country belongs.

But wait, you say to yourself. That's just marketing and advertising. It can't be that big a percentage of their budget, right? It's not like they spend more on advertising than research.

The CBS article has your answer:

> Yet nine out of 10 of the biggest pharmaceutical companies actually spend more on advertising that on R&D [Research and Development], according to The Washington Post. Sixteen drugs accounted for more than $100 million each in spending last year, with the most advertised drug being arthritis drug Humira, at $357 million . . .
>
> Among the most-advertised drugs is Jublia, one of the medications that bought time in the 2016 Super Bowl. The toe-fungus treatment costs about $600 a bottle but is proven to work in fewer than 20 percent of users, according to Consumer Reports.[5]

Nine of the top ten companies spend more on advertising than on research. Do away with advertising and you can double the research budget. How can anybody argue with that? Well, yeah, you might say, but advertising is advertising.

Everybody does it, right?

Nope.

The United States is one of only two countries in the world (the other is New Zealand) which allows direct to consumer advertising. In every other country, the pharmaceutical representatives have to convince the doctors that the drug is medically necessary and safe. In the United States and New Zealand, the patients come in demanding that new anti-fungal toe treatment or arthritis medication. We are being influenced in ways unfamiliar to the rest of the world:

> [T]he U.S. and New Zealand are the only two countries that allow pharmaceutical companies to market directly to consumers. Unfortunately for Americans, that doesn't appear to be helping health outcomes or spending. The U.S. spends more of its GDP on health care than 12 other developed countries.
>
> It may be no coincidence that citizens of only two countries take more than two prescription medications a day: The U.S. and New Zealand.[6]

I understand those who say correlation doesn't equal causation. But let's simply agree it's at least "suggestive" of a connection. If I eat a gallon of ice cream every night, then question why I can't fit into my favorite pair of jeans, I should at least give a thought to the ice cream.

And in talking about assumptions, remember that $5.2 billion dollars spent on pharmaceutical advertising in 2016? Well, the *Journal of the American Medical Association* did a study and found they'd underestimated the real number. The real number was closer to $6 billion dollars. They just missed it by $800 million dollars. *Forbes* did an article on the study, along with detailing what other investigations had found:

> According to the Journal of the American Medical Association (JAMA), there has been a drastic increase in DTC [Direct to Consumers) prescription drug marketing budgets, jumping from $1.3 billion in 1997 to more than $6 billion in 2016, that is a 361% increase. These companies use DTCA [Direct to Consumer

Advertising] to promote prescription drugs because it increases the volume of drugs sold. A 2008 House Commerce Committee found that for every $1,000 spent on prescription drug ads, 24 new patients were added for the pharma industry. Additionally, a 2003 research report found that sales for prescription drugs with ads were almost seven times greater than for those without ads.[7]

Those are certainly remarkable numbers. For every $1000 dollars spent on advertising, pharma picks up twenty-four new customers, for a per-patient cost of forty-two dollars. That means if you buy just one bottle of that Jublia antifungal for your toe at $600, they've got $558 dollars to divide up between the cost of production and other expenses.

And when Congress looked at the question of whether advertising works, the answer was an emphatic *yes*!

These advertising dollars are certainly buying influence with the public, but are they buying influence with the news stations that carry these ads? The answer is less clear. But I like to fall back on the old premise that the true golden rule is those who have the gold, make the rules.

It's not some evil conspiracy theory.

It's just the way the world works.

I encourage you to watch your favorite news station and count the number of pharmaceutical ads in a half hour block. It shouldn't take you long. Here's what the *New York Times* reported in 2017:

According to Kantar Media, a firm that tracks multimedia advertising, 771, 368 such ads were shown in 2016, the last full year for which data is available, an increase of almost 65 percent over 2012.

"TV ad spending by pharmaceutical companies has more than doubled in the past four years, making it the second-fastest-growing category on television during that time," Jon Swallen, Kantar's chief research officer, said.[8]

What the *New York Times* article didn't say is what percentage of the advertising budget would account for 771,368 ads in a single year. I'm guessing it's a big number, but it's difficult to find the information. Robert Kennedy Jr., who wrote the foreword to *Plague of Corruption*, does have a number, as he told the former governor of Minnesota, Jesse Ventura, in an interview in 2015:

> I ate breakfast last week with the president of a network news division [Roger Ailes of Fox News] and he told me that during non-election years, 70% of the advertising revenues for his news division come from pharmaceutical ads. And if you go on TV any night and watch the network news, you'll see they've become just a vehicle for selling pharmaceuticals. He also told me that he would fire a host who brought onto his station a guest who lost him a pharmaceutical account.[9]

That's about the best information you'll find on what percentage of the evening news is paid for by Big Pharma advertising. I'm not saying you should believe Kennedy. But why can't I even find any information to determine whether he's right or wrong?

* * *

What about Big Pharma and Congress? We can find information about how much the industry donates to Congress, and it will probably shock you. One of the best sources of information about the amount various industries spend lobbying Congress is a website called Open Secrets. But before I tell you the number, I want you to think of an industry that you think Congress isn't dealing with properly, and you suspect undue influence.

How about the oil and gas industry? That's one people like to talk about. Through the third quarter of 2019, the oil and gas industry shelled out a little over ninety-two million dollars. ($92,164,920.00)[10]

That's just a little more than thirty million a quarter. And who knows, maybe the oil and gas industry might want to give their favorite legislators a little bonus for Christmas to finish off the year?

Maybe you're not a fan of Big Tech and Silicon Valley. Perhaps you think they're throttling conservative voices, or other industries they don't like, such as mom and pop stores, so Target and Amazon can thrive. How much did Big Tech donate to Congress in the first three quarters of 2019? That number is a little over 119 million dollars (119,190,792.00).[11] That's just a shade under forty million dollars a quarter.

But now we come to pharmaceuticals/health products. They spend nearly double the amount as Big Tech, donating more than 228 million dollars in the first three quarters of 2019 ($228,149,734.00).[12]

Considering that they spend $6 billion dollars a year on advertising, Congress is a relative bargain.

If we assume their spending in the fourth quarter matches the first three, Big Pharma probably spent somewhere north of three hundred and four million dollars ($304,199,645) lobbying Congress in 2019, seventy-six million dollars per quarter ($76,049,911.00) or an average of more than eight hundred thousand dollars per day ($833,423.00).

Big Pharma is bigger than Big Tech and the oil and gas industry put together.

And with what we were saying about vaccines, we were messing with their profits. Now I've heard people say, "Yeah, Kent, I understand Big Pharma might be doing some bad things with their drugs, but they don't really make money on vaccines, do they?" According to the website *Statista*, the value of the global vaccine market in 2021 is estimated to be $47.7 billion and is estimated to grow to $61.6 billion by 2024.[13] (Personally, I think those numbers are very conservative, given our recent COVID-19 crisis.)

But there's one big difference between pharmaceuticals and vaccines.

Unlike pharmaceuticals, vaccines are a liability-free product, thanks to the National Childhood Vaccine Injury Act, as we previously discussed.

The children of our country, and indeed, all of us, are paying a very high price for President Reagan's mistake of signing the National Childhood Vaccine Injury Act. Others might shrink from the fight, but I will not.

Vaccines should live under the same consumer protection laws as every other product.

* * *

Although May 10, 2020 was when our book, *Plague of Corruption*, was the number one non-fiction book sold on Amazon for the week, it was clear even before that date that our book was screaming up the charts.

It was a threat to the Big Pharma narrative, especially with regards to the coronavirus story which had locked down the nation since mid-March. The same people who'd lied about vaccines and Judy's research were the ones everybody was looking to for information about the COVID-19 crisis.

Does Big Pharma get preferential treatment from the media for $6 billion a year and friends on both sides of the aisle for $300 million a year?

I suspect that for such an amount of money, it's not a "hope."

If you're going to give that kind of money, you probably need a guarantee.

On my side, I had a cartoonist, the wife of the former president of Fox News, Michelle Malkin, and a filmmaker named Mikki Willis who put together a twenty-six-minute video of *Plague of Corruption*. Was it any wonder the video was called *Plandemic*? Did that equal the $6 billion Big Pharma spent in the media or the $300 million a year they gave to Congress?

The media recipients of that $6 billion a year from Big Pharma needed to strike back. And they needed to do it fast.

First out of the gate was the *New York Times* with a story they published on May 9, 2020 with the title, "Virus Conspiracists Elevate a

New Champion,"[14] branding Dr. Mikovits and me as "disinformation agents." So, lest you think me a "disinformation agent," I quote directly from their article:

> Her ascent was powered not only by the YouTube video but also by a book that she published in April, "Plague of Corruption," which frames Dr. Mikovits as a truth-teller fighting deception in science. In recent weeks, she has become a darling of far-right publications like The Epoch Times and The Gateway Pundit. Mentions of her on social media and television have spiked to as high as 14,000 a day, according to the media insights company Zignal Labs.[15]

Okay, they've got six billion dollars they spend on advertising, and we get a few favorable mentions in *The Epoch Times* and *The Gateway Pundit*.

It still seems like Big Pharma has the overwhelming advantage.

And now we've got a "media insights company, Zignal Labs," discovering Judy was often getting fourteen thousand mentions a day on social media. What a threat to the republic she had become! And I want to know if somebody can tell me what a "media insights company" is, and who the hell pays them?

Dr. Mikovits and I are certainly not getting paid much. Since her fall from grace in 2011, she's never made more than fourteen thousand dollars a year, she lives on her husband's pension, and all they can afford is a five-hundred square foot studio apartment. But apparently, that hasn't stopped her from becoming an evil super-villain:

> The rise of Dr. Mikovits is the latest twist in the virus disinformation wars, which have swelled throughout the pandemic. Conspiracy theorists have used the uncertainty and fear around the disease to mint many villains. Those include Dr. Fauci after he appeared to slight President Trump and Mr. Gates, a co-founder

of Microsoft, as someone who started the disease. They have also pushed the baseless idea that 5G wireless waves can help cause the disease.[16]

While the *New York Times* did correctly report we'd been attacking Dr. Fauci online since 2018, I'd been attacking him in print since our 2014 book *Plague* was published.

However, Dr. Mikovits wasn't on social media until 2018.

And there were many sections of the *New York Times* article which were correct, such as:

> By April, coverage of Dr. Mikovits rose to 800 mentions a day. That month, Darla Shine, the wife of Bill Shine, a former Fox News executive and former top aide to Mr. Trump, promoted Dr. Mikovits's book in a tweet. Videos by *The Epoch Times*, a publication with ties to the Falun Gong, and the conservative outlet "*The Next News Network*" interviewed Dr. Mikovits about the pandemic, generating more than 1.5 million views on social networks.[17]

All of that is accurate. Darla Shine was an early supporter, as was *The Epoch Times* (a newspaper opposed to the communist party of China) and the *Next News Network*. Since I was also saying that both our government and the government of China had been experimenting with making bat coronaviruses able to infect human hosts, there was a question as to whether such experiments were the cause of this outbreak, which was killing the elderly and those with pre-existing conditions. The article continued:

> Then came the video from "Plandemic," which made mentions of Dr. Mikovits on social media spike far higher. The video was produced by Mikki Willis, who was involved in making "Bernie or Bust" and "Never Hillary" videos during the 2016 presidential campaign.[18]

Are you catching all of these players? The director of our twenty-six-minute video was a former "Bernie Bro," Mikki Willis, who was fighting the corruption of the democratic establishment as exemplified by Hillary Clinton.

Let's count up our supporters: the cartoonists Ben and Tina Garrison; the wife of the former head of *Fox News*, Darla Shine, who was a Democrat until 2016; Michelle Malkin, who seems to be a non-entity on Fox News these days; *The Epoch Times*, an anti-communist paper; the *Next News Network*; and Mikki Willis, a former Bernie Sanders supporter.

We truly were an island of misfit toys.

And I also have to give a shout-out to Zach Vorhies, better known as the "Google whistleblower." (Zach worked for YouTube, owned by Google.) Zach got me started on social media, and got attacked by the *New York Times* as well:

> Zach Vorhies, a former YouTube employee who has recently promoted QAnon conspiracy theories, posted a GoFundMe campaign on April 19 titled "Help me amplify Pharma Whistleblower Judy Mikovits." The campaign was first spotted by Ms. DiResta, of the Stanford Internet Observatory.
>
> A day before the GoFundMe campaign began, a newly created account for Dr. Mikovits tweeted for the first time. "A big thanks goes out to Zach Vorhies (@Perpetualmaniac) for helping me get on Twitter!" It was retweeted 400 times and liked more than 2,200 times. The account has gained over 111,000 followers in less than a month.[19]

As I've said, Judy is not much of a social media person, so it was great of Zach to help her out. I mean, who gets 110,000 Twitter followers in a month? Of course, Twitter then banned her from tweeting.

The *New York Times* was good enough to include a comment from our publisher. But of course they had to end the article by highlighting the danger I posed to society.

Skyhorse, the independent publishing company behind the book, defended its decision to print Dr. Mikovits. "The world should discuss the ideas in this book, rather than allow censorship to prevail," a spokeswoman for Skyhorse said.

Dr. Peter J. Hotez, dean of the National School of Tropical Medicine at Baylor College of Medicine, said her rise illustrated how the anti-vaccination movement had "taken a new ominous twist" with the coronavirus.

"They've now aligned themselves with far-right groups," Dr. Hoetz said, "and their weapons of choice are YouTube, Facebook and Amazon."[20]

What a remarkable idea from our publisher, to discuss the ideas in the book, rather than launching a personal attack on us. And how dare we use the formerly free speech platforms of YouTube, Facebook, and Amazon to communicate to the public? As of now, I'm not allowed on YouTube or Facebook, and our latest book, *The Case Against Masks*, was de-listed by Amazon after reaching #1 in the "Public Health" category.

I genuinely wonder how much longer we'll be a free country.

Two days after our book was #1 for the week on Amazon, on May 12, 2020, *Rolling Stone* published a long article with the title, "Judy Mikovits, Disgraced Doctor at the Center of "Plandemic," Has a Bestselling Book on Amazon: 'Plague of Corruption' is gaining traction on the Heels of a Proliferation of COVID-19 Conspiracy Theories."[21] Here's a taste of what they wrote:

Yet thanks to the notoriety of *Plandemic*, Mikovits' ideas have achieved mainstream visibility. She's gained more than 100,000 Twitter followers since April, and her 2020 book, *Plague of Corruption*, has skyrocketed to the bestseller list on Amazon, where it currently occupies the Number Seven slot (at one point, it had outsold the long-awaited follow-up to the *Twilight* franchise).

Indeed, according to a thread by Stanford Internet Observatory researcher Renee DiResta, there is evidence to suggest that Mikovits' appearance in *Plandemic* was actually part of an extensive marketing campaign for *Plague of Corruption*.[22]

Yeah, well, I've told you about that "extensive marketing campaign." And I certainly didn't spend $6 billion on it. I spent less than a couple thousand bucks. Okay, and who's this Stanford Internet Observatory egghead again who has evidence to suggest it was "part of an extensive marketing campaign"? Has the guy never heard of email? The phone? He probably could have figured out a way to contact us. He never did, though.

There were a couple other funny parts which read to me as if all those smart people spending their hundreds of millions and billions of dollars couldn't quite figure out how we'd pulled it off. I mean, they spend all this money to make sure people love Big Pharma: how did this story get past the guards?

Plague of Corruption, co-authored by anti-vaccine blogger Kent Heckenlively (and featuring a forward by noted anti-vaccine huckster Robert F. Kennedy Jr.), is, at first glance, an unlikely candidate for bestsellerdom. It's replete with scientific jargon largely impenetrable to the average reader, and with its countless references to internecine medical establishment grudges and squabbles, it reads more like an embittered relative's 10,000-word Facebook post against his former employer than a full-length book.[23]

When you read that section, isn't your first thought, *scientific jargon is the criticism*? In an age of jargon? *Flatten the curve? Spike protein? PCR test? Shedding? Social distancing? Mask up?*

Who is accusing whom of using too much scientific jargon? Over the past decade I've learned more than I ever wanted to know about persuasion and manipulation.

One of the common strategies to get a person worked up is to start with something true, and then go completely bonkers. If you get somebody to agree that the first thing you say is true, then they're not as primed to stop you when you veer off into crazy town.

I am a proud anti-vaccine blogger. You could also say I'm an attorney who worked for the Justice Department when I was in law school, the Rhodes Scholar candidate for my small liberal arts college, and a collegiate rower for three years. I worked for a United States senator in college, was the youth chairman for another candidate for the US Senate who'd been the police chief of Los Angeles, a fifteen-year science teacher, and with this book, have written or co-written ten published books. However, if *Rolling Stone* wants to call me an "anti-vaccine blogger," that's not an inaccurate statement.

But from me we move on to Robert Kennedy Jr., calling him an "anti-vaccine huckster." I'm not even sure what that's supposed to mean. If somebody wants to be a "huckster," it seems you'd want to go where the money is; like the nearly $50 billion a year Big Pharma makes from its liability-free vaccines, the $6 billion they spend on advertising, or the $300 million a year they spend lobbying Congress. Considering Kennedy seems to have chosen the approach that makes him the least money, by going after corporate giants like Monsanto and polluters of our air and water, "huckster" is the last label I'd attach to his name.

Now that *Rolling Stone* has the reader softened up, we get to the real insults.

The book is an "unlikely candidate for bestsellerdom," because it's "replete with scientific jargon largely impenetrable to the average reader" and to make matters worse, it has "countless references to internecine medical establishment grudges and squabbles." I don't know about you, but I had to look up the word "internecine."

I found it means a conflict destructive to both sides.

First of all, I think Judy and I did a great job of making the science understandable to the public. That's why the book has a 4.7 rating out of more than 6,500 Amazon customer reviews.

Second, in my opinion, *Rolling Stone* was trying to keep people away from the book by telling them, in effect, "It's too hard for you, the general public, to understand."

And third, they go on to say that even though it's "replete with scientific jargon," that "it reads more like an embittered relative's 10,000-word Facebook post against his former employer than a full-length book." The way I interpret that is that *Rolling Stone* is saying, "These are smart people who use big words that you won't understand. But they're also evil. So don't bother reading it."

Our book beat not only the former First Lady of the United States, but also those wimpy teen vampires and werewolves adored by adolescent girls across the planet. Let's move on to the final coup de grace of the *Rolling Stone* article:

> Yet Mikovits knows her audience, and she sets herself up as a courageous whistleblower raging against the machine rather than a disgraced scientist with a grudge against an establishment that rejected her.[24]

Are we really going to go with the "she's just an angry woman" argument? Is that the best they've got?

However, it didn't seem like the readers of *Rolling Stone* were going along with the new Big Pharma line taken by a magazine which had once been the vanguard of investigative journalism:

> Looks like Rolling Stone has been captured along with most government agencies, by those they are supposed to regulate. When you own the money, you own the media. When you own the media, you own the people, or at least the sheeple. Vaccines are a great deal because the cancer-causing mouse retroviruses are a free bonus. RIP Rolling Stone.[25]

Another reader made similar points:

> Thoroughly debunked? Where? I would like to see that! I have
> looked around, but have not found one of her points or positions
> responded to with data, or any kind of explanation. She, as well
> as dissenting voice, is attacked and vilified, but not one response
> or rebuttal. This is called the Ad Hominum Logical fallacy –
> personal attacks instead of arguing against the position or data.
> Couple that with censoring and we have no discussion. We have
> chosen experts telling us what to do.[26]

Perhaps you've heard of the claim about the media simply echoing each
other in what looks like some pre-arranged hit job. But until you've
been the subject of it, you really can't appreciate it.

On the same day as the *Rolling Stone* hit piece, May 12, 2020, there
was an article in *MedPage Today* with the title, "What Judy Mikovits
Gets Wrong: Pretty Much Everything, Pathologist Argues."[27] On May
14, 2020, *MedPage Today* posted another article with the title, "Point-by-
Point 'Plandemic' Smackdown: 'Dr. Mike Dissects the Viral Video."[28]

And we're the ones accused of cheapening science when they pub-
lish an article calling this debate a "smackdown"?

If it's such an open and shut case, why don't they just leave the
video up, so everybody can see what a fraud it is?

Then on May 16, 2020 there was the *New York Post* with an
article entitled "Who is Judy Mikovits, the Woman Behind Banned
'Plandemic' Conspiracy Video?"[29] Here's a sample from that article:

> Mikovits then turned to writing books with her co-author
> Heckenlively, churning out "Plague: One Scientist's Intrepid
> Search for the Truth about Human Retroviruses and Chronic
> Fatigue Syndrome (ME/CFS), Autism, and Other Diseases"
> in 2014 and in April of this year, the two released "Plague of
> Corruption: Restoring Faith in the Promise of Science"—now at
> number six on Amazon.

Two weeks later Mikovits dropped "Plandemic" on social media, in which she suggests there is a vast conspiracy to crush her career and destroy her credibility as a scientist. It also outlines her grudge match against Dr. Anthony Fauci and Dr. Deborah Birx, who are leading the White House Coronavirus Task Force.[30]

By the time this book is released we'll have been more than a year and a half into the COVID-19 emergency. How are you all feeling about the job done by Fauci and Birx? And was it really necessary to ban our twenty-six-minute video from all of social media? Am I that much of a threat to the public, or just to Big Pharma's nearly fifty-billion dollar a year liability-free vaccines?

And finally, on May 28, 2020, we got trashed by *Mother Jones* in an article titled, "After a Conspiracy Site Boosted the Debunked 'Plandemic' Video, Trump Pushed Its Take on Joe Scarborough."[31]

Don't believe the mainstream media works together?

In the space of a month we were featured in the *New York Times*, *Rolling Stone*, the *New York Post, MedPage Today,* twice, and *Mother Jones*, in addition to a number of other outlets which fed off the leftovers of the big guys like a pack of hyenas.

But the worst indignity was yet to come.

They needed to "talk about us" and yet "not talk about us" at the same time.

That mission was given to Chuck Todd of *Meet the Press,* and he executed it beautifully on his Sunday morning show of May 17, 2020.

TV had joined the print media in the attack on our books.

Big Pharma was going all in on their attack against us.

* * *

How do you talk about a book and not mention the title or the authors?

It sounds absurd when I say it, like something you might see in a Kafka book, or some fantasy tale like Harry Potter in which the villain,

Voldemort, is so feared that he's often referred to as simply, "He who must not be named."

Chuck Todd did his best to "talk" about us in a five-minute segment, but not really talk about us. Like Voldemort, we were "the book which must not be named."

> **Chuck Todd:** In the past three months, there's been a spike in belief in conspiracies, virtually all of them without factual basis, which makes them impossible to disprove. And that's the point. Joining me now is NBC News National Security Analyst Clint Watts, who specializes in tracking how conspiracies form and spread. He's a senior fellow at the Foreign Policy Research Institute and the Alliance for Securing Democracy.
>
> Clint, it's good to see you, good to see you're safe. Let me start with this. You have a way of mapping out these conspiracies and how they catch on. And you describe it this way: "How to build a disinformation bonfire." And in this case, like a bonfire, the first thing you need is a spark. And the spark being a theory is put online. In this case, why are these sparks so potent this time?[32]

The scene jumped from Chuck Todd at the *Meet the Press* studio, to Clint Watts, a serious-looking man in dark, horn-rimmed glasses, white airpods protruding from his ears, brown hair streaked with gray, blue jacket, white shirt, red tie, and behind him a colorful chart with three intersecting colored circles and titled, "The Year of Disinformation." Get it?

This is a serious man. He has a chart and he's going to tell you "the truth," because he's a fellow at the "Alliance for Securing Democracy." If there's a threat to democracy, this is the guy you want on the case. He's like James Bond with a PhD.

> **Clint Watts:** When we are scared, Chuck, in the time of pandemic, when we're not really sure what to believe, we're fearful both for our life and that of our families, we tend to take in information

that we might not otherwise consider. When you see these sparks thrown out there, there is a demand signal as well for it.

People want to know, "Are they safe? Can I trust this information?" And then people that have, you know, different agendas, maybe it's anti-vaxx, maybe it is reopening the country, maybe it's a conspiracy by elites. If you throw all those sparks out there at the same time, it can start a fire of disinformation.[33]

Chuck Todd nodded gravely, an image of the White House, the Washington Monument, and the reflecting pool as the banner for *Meet the Press* scrolled by like an old-time stock ticker. I also have to note that at one point in the next segment Todd looked nervous, as if he realized there were certain words (like our names or the title of our book), that he just couldn't say.

> **Chuck Todd:** Clint, what's amazing about this spark, and we're trying to be careful here not to give more attention to it. But there's this book and this movie that's circulated online. And while a bunch of tech companies got rid of it, the book itself is, you know, number one Amazon bestseller, number three *New York Times*. Both are putting it in the nonfiction category.[34]

Can't you just say the title of the book, Chuck? *Plague of Corruption.* Will your head explode if you do, or will your bosses fire you if you let slip the title or our names? And for God's sake, Chuck, a twenty-six-minute video is *not* a movie.

The screen returned to a glum-faced Clint Watts, defender of democracy. Seriously, if they're going to trash us, shouldn't we at least get a couple minutes of air time? In this book, I'm nice enough to let them speak for themselves, and let you make up your own mind.

> **Clint Watts:** Chuck, this is a super organized effort. You've got multiplatform, multimedia. You've got people who know how to

push this stuff in the social media environment. You have people who know how to make a very high-production video that's
very engaging. It makes you want to start to engage in these conspiracies. And then you've got the so-called background. This is
that book. This is many, many websites that are out there talking
about health or health risk relating to vaccines.[35]

This guy Clint is supposed to be so smart, but he could have gone to
our website, www.plaguethebook.com, and left a message. Didn't he
at least take a moment to consider we might have a website where he
could contact us? We would have responded to him. Maybe we could
have informed him about the Vaccine Court. I suspect he doesn't know
that pharmaceutical companies have absolutely no liability for the injuries or deaths caused by their vaccines.

Chuck and Clint talked a little more about the "kindling" necessary to create a "bonfire of disinformation," before Todd led into his
final points.

Chuck Todd: And four years ago, this disinformation campaign
started internationally, and then domestic actors picked it up. It
appears this is just the opposite this time, right? This is, we're
doing this to ourselves, but the international folks are goosing it
a little bit, aren't they?[36]

There it is. We're helping the Russians and that evil Vladimir Putin.
Like Sherlock Holmes blaming everything on the sinister Dr. Moriarty,
the mainstream media tries to tie everything they don't like in America,
including us, to the Russians.

Hey, and while we're at it, why don't we throw in the Chinese
and the Iranians? Although it's a little difficult to figure out why an
anti-Chinese communist paper like *Epoch Times* was giving us such
favorable coverage.

But hey, when you've got a good story, who cares about consistency?

Clint Watts: Chuck, when I talked to you three years ago about disinformation coming from Russia, it was something they had to push along. They had to pick and grab things in the U.S., but they also made their own conspiracies to push. That's no longer the case. Right now, whether it's Russia, Iran, China, or any other adversary, there's plenty of U.S.-made disinformation which they can pick and choose, amplify, and send back into the U.S. audience's face, seeking to divide us, seeking to erode trust in institutions, seeking to erode trust in our health care system, which is something that's essential at this point.[37]

Do you realize how malevolent this propaganda is for our country? On one level, it's screamingly funny and absurd, which is why I mercilessly mock the mainstream media when they make these attacks.

But on another level, it's absolutely terrifying.

They're trying to convince our fellow citizens that Judy and I are working for an unfriendly foreign power. The mainstream media wants to make their audience believe we're part of a right-wing conspiracy. But here's the truth.

The left-wing media viciously attacks us.

The right-wing media, like *Fox News* and *Brietbart*, act like we don't exist, despite our tremendous success.

It seems like one side of the aisle has been paid to be the attack dog and the other side has been paid to look the other way.

Our true friends are the corruption fighters, whether they are like Darla Shine and Michelle Malkin on the right, or Mikki Willis and Robert Kennedy Jr. on the left. It appears the wolves have got the sheep convinced the wolves are the good guys, and the shepherds like us, watching out for the flock, are the bad guys.

That's the magic trick six billion dollars a year in advertising can buy you.

* * *

If you change the laws, as with the establishment of the "Vaccine Court," then use government power to force a medical product upon the people and pay off the media to attack any critics of your plans, is it any wonder that people do not trust the vaccine mandates? As Malcom X had warned decades ago, "If you're not careful, the newspapers will have you hating the people who are being oppressed and loving the people who are doing the oppressing."[38]

CHAPTER EIGHT

My Body Is My Property

When I began this book, I believed the main point would be simple: My body is my property.

I'm not a vegetarian, but if you were a vegetarian and came to my home for dinner I wouldn't make you eat a steak.

I do not expect any government to tell an Orthodox Jew or Muslim they must eat pork.

I'm not one of those people who tells you who you must vote for, the newspapers, magazines, or cable news channels you must follow, or the views you must hold.

However, I have no hesitation in telling you what I think.

But if you think differently from me, more power to you. I'm more than happy to sit down and share a coffee with you to chat about the world. Maybe you'll convince me of your point of view.

I'm open.

For me, that's just the regular functioning of a civilized society.

When my ninety-one-year-old father was in failing health in April 2021 and asked my opinion as to whether he should continue with kidney dialysis, I told him it was his decision. I would respect any decision he made. "I just don't see an end to it," he told me. "It all seems like a scam."

I never wanted to be one of those adult children who told an elderly parent, especially one like my father who had all of his cognitive faculties, what he could and could not do. I suggested a middle course, maybe just one or two treatments a week, then seeing if it was any different. He did that for a few weeks; then when I went down to see him in Palm Springs again, he told me it was just as bad.

He made the decision to stop the dialysis treatments and died in early July 2021. I do not know if there is any such thing as a "good death," but it happened on his own terms. He was a free and independent person, in charge of his own health, without force or compulsion, until the day he died.

I miss him greatly and hope my family will treat me with similar respect when it is my time.

* * *

A genuine surprise to me in the writing of this book was to discover how flimsy the legal precedence is for mandatory vaccination.

I had not expected the 1905 *Jacobson v. Massachusetts* case, which all the legal experts seemed to cite as controlling law on this issue, to be so easily distinguishable on the facts. Pastor Jacobson never had to take a smallpox vaccination and was only fined five dollars for his refusal to submit to the shot. Nothing I've been able to uncover suggests he was unable to travel on a ship, have a job, or interact in civil society, as many of our political leaders are suggesting today for those unwilling to take a COVID-19 vaccine.

I was shocked to discover the hidden power of the eugenicists from the late 1870s to the outbreak of World War II, and learn that their ranks included prominent members of society, such as Oliver Wendell Holmes Jr., who wrote the odious decision in the *Buck v. Bell* case in 1927. That case equated the imposition of a fine for refusing a vaccination to the involuntary cutting of the Fallopian tubes of a mental patient, a procedure which was eventually inflicted on more than sixty thousand Americans. As a conservative, I have generally disagreed with the positions of progressives. But nothing prepared me for the shock of learning that Harvard and Stanford Universities were the academic center of the eugenics movement in America, or that eugenics research was, for many years, the number one focus of the famed Cold Springs Harbor Laboratory, one of our great national laboratories.

Prior to writing this book I did have some familiarity with the Nazi program to kill disabled children, but did not realize how much of their twisted intellectual framework was based on the work of American eugenicists and their sponsors, such as the Rockefeller and Carnegie Foundations. And while I was aware of the Doctors' Trial at Nuremberg after World War II, I did not know how many of those who worked in these programs were able to quickly find their place in the postwar world without any consequences for their actions.

I'm aware that many who have concerns about vaccines believe the vaccine program itself is part of a modern-day eugenics program, led by Bill Gates, the World Health Organization, and other groups.

I can find no evidence to support this view.

And yet, at the same time I'm troubled by the fact I can find no former American eugenicists who were either prosecuted or publicly renounced their beliefs. There was no purge of eugenicists at Harvard, Stanford, Cold Springs Harbor, or at the Rockefeller or Carnegie Foundations.

It is as if they simply vanished.

Make of that what you will.

* * *

The second part of the book, in which I went over the history of the vaccine program, was more familiar to me.

I'm sure that for many it will be the first time they're reading about the 1986 National Childhood Vaccine Injury Act (NCVIA) or the Vaccine Injury Compensation Program (VICP), as well as what academics and those who have worked in it have to say about the program.

They say that terrible things come in threes, and I can think of nothing more horrible than the revelations of the Simpsonwood Conference of June 6–7, 2000, the revelations of senior CDC scientist Dr. William Thompson about the effect of earlier administration of the MMR shot on Black boys in 2014, or what Dr. Andrew Zimmerman, the government's main scientific witness in the Autism Omnibus Proceeding, revealed in 2019, namely that in June 2007 he was ready to testify that at least a third of the autism cases were due to vaccine injuries.

When Dr. Mikovits and I published our second book, *Plague of Corruption,* in April 2020, in the midst of the COVID-19 crisis, I wish I could say I was surprised by the negative reaction of the mainstream media.

But I wasn't.

I'd seen the same song and dance used against Dr. Andrew Wakefield and others. The playbook wasn't surprising to me. There was no reason to get angry. When you're bucking an industry that makes tens of billions of dollars a year and has no legal liability, they're going to fight back.

Frankly, I'm surprised I'm still alive and free to walk the earth.

And as I believe I have an eternal soul and what I do matters, perhaps the next obvious question is: What do we do next?

* * *

If you're a person of faith, part of the answer is believing God will move other people to the places where they need to be in order to accomplish His plans. As for me, I see signs of that everywhere.

On August 22, 2021, the *New York Times* published an article with the title, "A Hospital Finds an Unlikely Group Opposing Vaccination: Its Workers."[1] From the opening of the article:

Their movement started discreetly, just a handful of people communicating on encrypted apps like WhatsApp and Signal. But in just days it had ballooned tenfold. And within two weeks, it had turned into a full-blown public protest, with people waving picket signs to denounce efforts to push them to receive coronavirus vaccines.

But these were not just any vaccine resisters. They were nurses, medical technicians, infection control officers and other staff who work at a hospital on Staten Island, which had the highest rate of Covid-19 infection of any borough in New York City.

Outside Staten Island University Hospital this week, as passing cars and fire trucks honked supportively, employees chanted, "I am not a lab rat!"[2]

To me this sounds like the beginning of a full-blown resistance to vaccine mandates and the pressure campaign which, in addition to the loss of employment, includes a healthy dose of social shaming. You can't be shamed for being overweight or for whom you love, but you can be shamed if you decline a vaccine for which the companies which produce them have no legal liability.

And how big of a potential problem is this among health-care workers and what are the modes of persuasion that powerful people are using to convince these resisters that the vaccines are safe?

Across New York, the majority of the state's more than 600,000 health care workers are vaccinated, but many are not. To date, 75 percent of the state's roughly 450,000 hospital workers, 74 percent of the state's 30,000 adult care facility workers, and 68 percent of the state's 145,000 nursing home workers have been fully vaccinated, the state said.

Modes of persuasion ranging from free cash to burgers to rides on the M.T.A. failed to persuade vaccine refusers, leading some hospital systems to take a harsher approach, which in turn, has spurred a backlash. Last month, the largest health care union

in the country held a rally after the New York-Presbyterian hospital system mandated that workers receive at least one shot of the
vaccine by Sept. 1.[3]

When one reads about these efforts, the inescapable conclusion is that a
significant percentage of our most scientifically-literate population has
made the decision that the COVID-19 vaccines are not safe, and they
will not take them. As I read the response of those in power, they have
decided not to engage in a substantive conversation. They have elected
instead to treat the vaccine refusers as children, offering them money
like a grandparent might do with a grandchild for good grades, or saying that if they're good and take the shot, they'll get a free cheeseburger
and maybe a shake.

I genuinely wish the other side would engage in a conversation with
us about our concerns.

Maybe they believe they are.

The article further detailed how the leader of the vaccine refusers
at Staten Island University Hospital, John Matland, a CT scan techni-
cian, took part in a survey put forth by the hospital administrators as
to the possible steps which might motivate the staff to be vaccinated.

> Mr. Matland participated in the focus group aimed at under-
> standing what punitive measures could motivate unvaccinated
> employees to get the shots. The options listed were: being docked
> pay during leaves of absence if exposure requires quarantine;
> becoming ineligible to participate in employee appreciation bar-
> becues; or losing points that staff are allowed to cash in for gift
> cards and products.
>
> Mr. Matland said he chose "none of the above."[4]

The approach is so clumsy one half expects it isn't real, but rather something from a *Dilbert* comic strip, in which the main character is for-
ever pointing out that corporate management is always choosing the

stupidest of all possible options. Oh, no! I can't go to the employee appreciation barbecue? My life is ruined! I won't be allowed to cash in my employee points for gift cards and products like a T-shirt emblazoned with the name of my corporate masters? Oh, the humanity!

Even among those health-care workers who took the shot, there seems to be a reluctance to enforce vaccine mandates on their fellow workers, as demonstrated by this article about a California health-care worker:

> "The principle of forcing a vaccine on someone goes against everything we've ever been taught as nurses," Modesto nurse Laura Estrella said.
>
> Estrella said she wants the same rights as her patients.
>
> "Obviously, we've all taken our vaccines. In order to get into the nursing program, you have to have vaccines. And then there's the yearly flu recommended vaccine, people take these too," she said. "We're not against those. We just don't want to be forced. That's where it crosses the line."[5]

You might need to forgive me for saying so, but one of the bedrock principles of American life is you get to make your own decisions about your life. Employers have a limited ability to pry into your personal life or medical details. And yet now, with these vaccine mandates, your private decisions are open for the world to see.

These protests are even happening outside of the more populous states, as demonstrated by this article about a nurse's protest in Nebraska:

> Wynee Benedict is a registered nurse. She says she wants to be the voice for people who aren't as comfortable speaking out on the vaccine mandate.
>
> "I stood up for my community. I stood tall and proud during a time of uncertainty when a lot of people were scared. A lot of us stood tall. We stood strong for our patients, for our communities. And now we feel like we're being let down," said Benedict.[6]

This Nebraska nurse feels like a lot of health-care workers. They worked on the front lines during the COVID-19 crisis, and now they're being singled out for coercion.

The health care they learned was never about coercion. It was about educating and allowing the patient to make an informed decision.

It is perhaps not surprising that the most principled voices are coming from the ranks of our public servants, such as this teacher who supported Wynee Benedict's protest:

> Diane is protesting today in support of a local nurse in Nebraska City. She's concerned that the requirements will be bigger than just health care workers.
>
> "As a teacher, I'm worried that if that trend continues then school employees, teachers, para's, possibly students would be the next ones that would be forced," said Diane . . .
>
> "That's stuff I am already looking into. I will not quit my job if I am forced to, I would make them terminate me. I've already decided that," said Diane.[7]

Diane is correct to be concerned. They are coming after all of us. The budding tyrant lurking in the souls of many public officials is showing itself and we must resist. It is interesting to note that those among us who are the greatest sustainers of civilization—medical professionals, teachers, and police—seem to understand this threat much earlier than the rest of the public.

This is from an article about the response of police in Chicago to a possible COVID-19 mandate:

> After Chicago Mayor Lori Lightfoot said all city employees would "absolutely" be required to be vaccinated, the leading police union vowed to take action in court if such mandate is announced.
>
> "We are 100% against mandated vaccines for our members," Chicago Fraternal Order of Police president John Catanzara,

whose union represents rank-and-file police officers in a statement to Fox 32.

"This vaccine has no studies for long-term side effect or consequences. None," he continued in a similar statement to the Chicago Sun Times on Tuesday. "To mandate anybody to get that vaccine, without that data as a baseline, amongst other issues, is a 'Hell, no' for us."[8]

Nurses, teachers, and cops (in addition to white, well-educated Whole Foods moms, Blacks, and Hispanics) are all joining together to fight against COVID-19 vaccine mandates. But the fight against vaccine mandates is gathering some truly unprecedented alliances. This movement is spreading across the globe, as detailed in this August 2021 article from *Voice of America*:

> More than 230,000 people marched Saturday in cities across France to protest measures meant to counter the spread of the coronavirus, including vaccines for health care workers and a health pass needed to access many public indoor activities.
>
> The fourth week of protests was also the largest and included marches on Paris, Nice, Montpellier and Lyon, where police used tear gas on protestors who threw objects at police. The measures have brought together France's hard-left anarchists and hard-right militants, according to Reuters.
>
> Health care workers have until September 15 to get their shots or face suspension.[9]

The French, who value freedom so much that they sent us our own Statue of Liberty to commemorate our common love of the God-given rights of human beings to make their own decisions, are standing up to this tyranny as well.

* * *

Perhaps the greatest surprise is those vaccine refusers who are least likely to change their minds are not necessarily nurses, cops, or teachers, but those with the highest level of education. As described in an article from Britain's *Daily Mail*:

> People with a PhD are the most hesitant when it comes to getting the Covid-19 vaccine, according to a paper by researchers from Carnegie-Mellon University and the University of Pittsburgh.
>
> Researchers surveyed just over five million US adults in an online survey, with 10,000 reporting they were educated to PhD level.
>
> The report showed a surprising U-shaped correlation between willingness to get a Covid-19 vaccine and education level—with the highest hesitancy among those least and most educated.
>
> Of those surveyed, 20.8 percent with a high school education were reluctant to get the shot, and 23.9 percent with a PhD were against it.
>
> But the least skeptical of the shot had a Master's degree, with only 8.3 percent of that group being vaccine hesitant.[10]

It can be a challenge to make sense of these findings, but I have an explanation based on my own experiences. As an attorney who is now an educator, I'd say I have the greatest amount of education relative to my fellow teachers, who generally have a master's degree. After the numerous controversial books I've published, one might expect that would create some curiosity about my writing.

That's how you imagine educators would act, right? "Oh, you wrote a book? I'd love to read it, then maybe discuss some of your ideas over coffee. Can you give me a couple weeks to read it and collect my thoughts?"

That has not happened once in the past ten years. I work in a school where there are posters which proclaim, "Everybody is welcome!" and "Diversity is our strength!" but I am a pariah.

Except for the custodians.

The custodians, who come into my room near the end of the day to clean, are fascinated by what I write. I give every one of them free copies of my books and they eagerly devour them, anxious to come to my classroom to discuss the ideas. Maybe the commonality is that when you're on the bottom of the economic ladder you realize things are screwed up. But as you're climbing the ladder, you believe in the system. If that's where you stop, you think everything is great and your leaders are wise and well-meaning.

But maybe when you get to the top of the educational ladder, you survey the scene and say to yourself, "Yeah, everything is just as screwed up as I thought it was. Those in charge have no idea what they're doing."

As for me, I have made my own calculations. Although I do not believe I am at risk, I have a supply of hydroxychloroquine and ivermectin in my medicine cabinet, purchased legally through America's Frontline Doctors, an organization I encourage you to follow at www.americasfrontlinedoctors.org. This is just my opinion about the smart thing to do in a very complicated world. There are other therapies on the horizon, such as Regeneron, currently being used in Texas and Florida, which in a phase III trial in March 2021 showed it "significantly reduced the risk of hospitalization or death by 70%."[11]

When one researches other therapies to treat COVID-19, the impression left is that many in the medical community are like a workman who has only a hammer in his tool chest, and thus sees every problem as a nail. There are many ways to address a problem. Science is about having an open mind and realizing that good health does not come from the tip of a needle. Instead, it comes from healthy habits, remaining open to possibilities, and a healthy skepticism of those who tell you to simply shut up and take your place in line.

And so, what is to be done?

Well, a lot of this is happening without any coordination because a good number of the public is awake and realizing that the tyrants are coming for our jobs and our ability to interact in society. Therefore, I encourage you to join whatever groups you can, or support those who are in this fight.

As Benjamin Franklin said to his fellow patriots at the beginning of our country's war of independence in 1776, "We must indeed, all hang together or, most assuredly, we shall all hang separately."

What Comes Next

When I first met my lovely wife, Linda, in 1992, I quickly noticed something special about her. She cared about people and genuinely wanted to help them through her job as a speech therapist.

I know when most people think of a speech therapist they picture somebody who works with children. My wife works with adults, generally stroke victims, accident victims, people with neurological diseases, or those who've recently undergone some type of surgery which has affected their ability to swallow, eat, or talk. "I like the ones who come in mad," she told me early on. "They're pissed because their life has been suddenly turned upside down, and they don't know if it's ever going to get better. Give me forty-five minutes with that person and I'll turn them around."

My wife is like the police or soldiers who run toward the sound of gunfire, rather than away. And when she comes home at night, the first half hour is usually her telling me about what happened with her patients that day.

Please understand I am the opposite. I generally run away from conflict. I like to spend my days with happy people. I did not like being an attorney because my job in effect was to ruin the other people's days. A successful attorney is one who gets big money for his clients, or prevents his client from having to pay big money.

I made a mid-life switch to teaching science fifteen years ago and have never looked back. My job these days is to convince students they *can* do something, rather than making somebody *pay* for what they did, or failed to do.

I like to be on the positive side of the human experience.

My days are filled with nothing more controversial than teaching the theory of continental drift, the organization of a typical cell, and how energy might move through an ecosystem. I laugh a lot, tell jokes, and people always ask why I'm so upbeat. The answer is it's simply my nature. I seem to have a talent for happiness, even in difficult situations.

I'm aware most people know me through my writing, which is critical of many large institutions, such as Big Pharma and Big Tech. I consider it my duty as a citizen to fight against injustice in any way I can, and I am proud of what I have done. You might say my side-hustle is fighting bad guys, so in some sense, like my wife, I also run toward the sound of gunfire.

However, on September 30, 2021, my wife will be terminated from her job as a speech therapist after more than thirty-three years in the profession because she will not take the COVID-19 vaccine.

My job as a middle school science teacher is also uncertain. On August 11, 2021, California Governor Gavin Newsom announced the rules that will apply to teachers in the state:

> California will become the first state in the nation to require all teachers and staff to get vaccinated or undergo weekly COVID-19 testing, as schools return from summer break amid growing concerns about the highly contagious delta variant, Gov. Gavin Newsom announced Wednesday.
>
> The new policy applies to both public and private schools and will affect more than 800,000 employees, including about 320,000 public school teachers and a host of support staff such as cafeteria workers and cleaners, the state Department of Public Health said. It will also apply to school volunteers.[1]

I will also not take any COVID-19 vaccine, for many of the reasons I have listed in this book. I suspect the teachers are being softened up by this approach and that by the end of the year I will be given the option of taking a vaccine or being fired.

We need to not only resist the current mandatory vaccination plans, first put into motion by the eugenicists of the early nineteenth century and handed off to the tyrants of today, but also we need vaccines to be investigated under the same laws as every other consumer product.

Humanity is under no greater threat than that of mandatory, uninvestigated vaccines, and their harmful side effects. Future generations will look back on us and ask if we rose to meet this challenge.

Vaccine mandates will usher in a new dark age of terror in which your body is merely the property of somebody else. This is the same thinking which supported slavery, eugenics, and Nazi tyranny.

We must resist, even if we are driven from our jobs, places of public gathering, and who knows what may come after that. It is time to stand up and say this is a line no government or corporation will be allowed to cross.

Endnotes

Chapter One

1. Helen Elfer, "Laura Ingraham Loses Her Cool as Alan Dershowitz Challenges her Anti-Vax Narrative," *Yahoo Money*, July 30, 2021, www .sports.yahoo.com/laura-ingraham-loses-her-cool-154129483.html.
2. Ibid.
3. Ibid.
4. *Jacobson v. Massachusetts*, 197 U.S. 11 (1905).
5. Preamble to the United States Constitution, www.archives.gov/founding -docs/constitution-transcript (Accessed August 6, 2021).
6. Fourteenth Amendment to the US Constitution, approved by Congress on June 13, 1866, and ratified on July 9, 1868, www.constitution.congress .gov/browse/amendment-14/ (Accessed August 6, 2021).
7. *Jacobson v. Massachusetts*, 197 U.S. 11, p. 12 -14 (1905).
8. Wendy k. Mariner, George J. Annas, and Leonard H. Glantz, "Jacobson v. Massachusetts: It's Not Your Great-Great Grandfather's Public Health Law," *American Journal of Public Health*, vol. 95(4), p. 581–590 (April 2005), doi: 10.2105/AJPH.2004.055160.
9. "Smallpox," American Museum of Natural History (Accessed August 6, 2021), www.amnh.org/explore/science-topics/disease-eradication/countdown -to-zero/smallpox#:~:text=Lessons%20from%20the%20Past&text=One %20of%20history's%20deadliest%20diseases,million%20people%20 since%201900%20alone.
10. Wendy K. Mariner, George J. Annas, and Leonard H. Glantz, "Jacobson v. Massachusetts: It's Not Your Great-Great Grandfather's Public Health Law," *American Journal of Public Health*, vol. 95(4), p. 581–590 (April 2005), doi: 10.2105/AJPH.2004.055160.

Chapter Two

1. *Buck v. Bell, Superintendent of State Colony Epileptics and Feeble Minded*, 274 U.S. 200 (1927).
2. Ibid.
3. Barbara Loe Fischer, "Forced Vaccination: The Tragic Legacy of Jacobson v. Massachusetts," National Vaccine Information Center, November 2, 2016, www.nvic.org/NVIC-Vaccine-News/November-2016/forced-vaccination-the-tragic-legacy.aspx.
4. Alexandra Minna Stern, "That Time the United States Sterilized 60,000 of Its Citizens," *Huffington Post*, January 1, 2016, www.huffpost.com/entry/sterilization-united-states_n_568f35f2e4b0c8beacf68713.
5. Ibid.
6. Kara Goldfarb, "21 Historical Figures You Didn't Know Supported the Eugenics Movement," *All That's Interesting*, July 19, 2018, www.allthatsinteresting.com/eugenics-movement.
7. Ibid.
8. Adam S. Cohen, "Harvard's Eugenics Era," *Harvard Magazine*, March-April 2016, www.harvardmagazine.com/2016/03/harvards-eugenics-era .
9. Ibid.
10. Ibid.
11. Ibid.
12. Ibid.
13. Edwin Black, "The Horrifying Roots of Nazi Eugenics," *History News Network*, September 2003, www.historynewsnetwork.org/article/1796.
14. Cold Spring Harbor Laboratory Website, Front Page, (Accessed August 7, 2021), www.cshl.edu/
15. Joshua A. Krisch, "When Racism was Science," *New York Times*, October 13, 2014, www.nytimes.com/2014/10/14/science/haunted-files-the-eugenics-record-office-recreates-a-dark-time-in-a-laboratorys-past.html.
16. Ibid.
17. Anna Diamond, "The 1924 Law that Slammed the Door on Immigrants and the Politicians Who Pushed it Back Open," *Smithsonian Magazine*, May 19, 2020, www.smithsonianmag.com/history/1924-law-slammed-door-immigrants-and-politicians-who-pushed-it-back-open-180974910/.
18. Edwin Black, "The Horrifying Roots of Nazi Eugenics," *History News Network*, September 2003, www.historynewsnetwork.org/article/1796.
19. Hannah Ward Glenton, "Companies Making Big Money from COVID-19 Vaccines," *Microsoft News*, April 5, 2021, www.msn.com/en-us/money/news/companies-making-big-money-from-covid-19-vaccines/ss-BB1faZ9R#image=1.

20. Alfred Lubrano, "Anti-Vaccine Parents are Often White, College-Educated, 'Whole Food Moms," *Philadelphia Inquirer,* April 10, 2019, www.inquirer .com/news/middle-class-working-class-vaccine-anti-vaxxers-measles-cdc -20190410.html.

21. Ibid.

22. Ibid.

23. Barbara Feder Ostov, "State Broadens Investigation of Doctors For Issuing Questionable Vaccination Exemption," *California Health* online, July 2, 2019, www.californiahealthline.org/news/state-broadens-investigation-of-doctors-for-issuing-questionable-vaccination-exemptions/

24. Adam Beam, "California to Mandate COVID-19 Vaccines for Health Workers," *U.S. News and World Report,* August 5, 2021, www.usnews.com /news/politics/articles/2021-08-05/california-to-mandate-covid-19-vaccines -for-health-workers.

25. John Bowden, "Major Teachers' Union Head Endorses Vaccine Mandate for Educators," *Yahoo News,* August 8, 2021, www.news.yahoo.com/major -teachers-union-head-endorses-180139258.html

26. Joshua Rhett Miller, "Boston Mayor Compares NYC's Vaccine Mandate to Slavery, Birtherism," *New York Post,* August 4, 2021, www.nypost .com/2021/08/04/boston-mayor-compares-nycs-vaccine-mandate-to-slavery/

27. Adrian Florido, "Many Hispanic and Black People Say They Are Skeptical about COVID-19 Vaccines, " NPR, December 10, 2020, www.npr.org /2020/12/10/945136592/many-hispanic-and-black-people-say-they-are -skeptical-about-covid-19-vaccine

28. Wendell Huesbo, "Dr. Drew: Vaccine Passports are 'Modern Day Segregation,'" *Breitbart News,* August 4, 2021, www.breitbart.com/politics /2021/08/04/dr-drew-vaccine-passports-are-modern-day-segregation/

29. Patrick Reilly, "SF Sheriff's Deputies Threaten to Quit over COVID-19 Vaccine Mandate," *New York Post,* August 7, 2021, www.nypost.com /2021/08/07/san-francisco-sheriffs-vow-to-quit-if-forced-to-get-covid-19- vaccine/.

30. Carolyn Abraham, "Vioxx Took Deadly Toll: Study," *The Globe and Mail,* January 25, 2005, www.theglobeandmail.com/life/vioxx-took -deadly-toll-study/article1113848/.

31. Luke Timmerman, "Failing to Report Severe Drug Side Effects: A National Embarrassment," *Forbes,* May 1, 2015, www.forbes.com/sites /luketimmerman/2015/05/01/under-reporting-of-severe-drug-side-effects-a -national-embar rassment/?sh=434a222b251d

32. Megan Redshaw, "VAERS Latest Data Include 2 New Reports of Teen Deaths Following COVID Vaccine, as Total Reports of Death Exceed

12,000," *The Defender*, August 6, 2021, www.childrenshealthdefense.org /defender/vaers-cdc-data-injuries-deaths-covid-vaccine/.

Chapter Three

1. "The Doctors Trial: The Medical Case of the Subsequent Nuremberg Proceedings," United States Holocaust Memorial Museum, (Accessed August 9, 2021), www.ushmm.org/information/exhibitions/online -exhibitions/special-focus/doctors-trial.

2. "Euthanasia Program and Aktion 4," United States Holocaust Memorial Museum, (Accessed August 9, 2021), www.encyclopedia.ushmm.org /content/en/article/euthanasia-program.

3. Ibid.

4. Ibid.

5. Ibid.

6. Ibid.

7. "The Doctors Trial: The Medical Case of the Subsequent Nuremberg Proceedings," United States Holocaust Memorial Museum, (Accessed August 9, 2021), www.ushmm.org/information/exhibitions/online -exhibitions/special-focus/doctors-trial.

8. Nazi Medical Experiments: Background and Overview: Background and Overview," Jewish Virtual Library, (Accessed August 10, 2021), www .jewish-virtuallibrary.org/background-and-overview-of-nazi-medical -experiments.

9. Peter Tyson, "Holocaust on Trial: The Experiments, NOVA Online, (October 2000), www.pbs.org/wgbh/nova/holocaust/experiside.html.

10. "Josef Mengele, Known as the 'Angel of Death,' Dies," *History Channel*, November 5, 2009, updated February 4, 2021, www.history.com/this-day -in-history/the-angel-of-death-dies.

11. Edwin Black, "The Horrifying Roots of Nazi Eugenics," *History News Network*, September 2003, www.historynewsnetwork.org/article/1796.

12. "The Nuremberg Code (1947)," *British Medical Journal*, v. 313, 1448, (December 7, 1996) doi.org/10.1136/bmj.313.7070.1448.

13. Ibid.

14. Hana Weingarti, Markus Czub, Stephanie Czub, et. al., "Immunization with Modified Vaccinia Virus Ankara-Based Recombinant Vaccine against Severe Acute Respiratory Syndrome is Associated with Enhanced Hepatitis in Ferrets," *Journal of Virology*, Vol. 78(22), p. 12672–12676, doi. org/10.1128/JVI.78.22.12673–12676.2004.

15. Wen Shi Lee, Adam K. Wheatley, Stephen J. Kent, and Brandon J. DeKosky, "Antibody-Dependent Enhancement and SARS-CoV-2 Vaccines

and Therapies," *Nature Microbiology,* 10, p. 1185–1191, doi: 10.1038/s41564-020-00789-5.

16. Stephanie Seneff and Greg Nigh, "Worse than the Disease? Reviewing Some Possible Unintended Consequences of the mRNA Vaccines Against COVID-19," *International Journal of Vaccine Theory, Practice, and Research*, Vol, 2, No. 1, May 10, 2021, www.ijvtpr.com/index.php/IJVTPR/article/view/23.

17. Harriet Washington, "Non-Consenting Adults," *Slate,* January 22, 2012, www.slate.com/technology/2012/01/the-nuremberg-code-set-up-to-protect-the-human-subjects-of-research-is-being-routinely-ignored.html.

18. Ibid.

19. Ibid.

Chapter Four

1. Nora Freeman Engstrom, "A Dose of Reality for Specialized Courts: Lessons from the VICP," *University of Pennsylvania Law Review*, Vol. 163, 1631–1717, 1655 (2015).

2. Ibid. at 1656.

3. Ibid.

4. Ibid at 1657.

5. Robert Pear, "Reagan Signs Bill on Drug Exports and Payment for Vaccine Injuries," *New York Times*, November 15, 1986, www.nytimes.com/1986/11/15/us/reagan-signs-bill-on-drug-exports-and-payment-for-vaccine-injuries.html.

6. Telephone Interview with Nora Freeman Engstrom by Kent Heckenlively, February 18, 2016.

7. Ibid.

8. Telephone Interview with former Chief Special Master Gary Golkiwiecz, January 29, 2016.

9. Ibid.

10. Ibid.

11. Ibid.

12. Ibid.

13. Ibid.

14. Ibid.

15. Ibid.

16. Jennifer Schlesinger and Karina Hernandez, "Compensation for COVID Vaccine Injuries is Limited," *CNBC*, March 25, 2021, www.cnbc.com/2021/03/25/compensation-for-victims-of-covid-vaccine-injuries-is-limited.html.

17. Luke Timmerman, "Failing to Report Severe Drug Side Effects: A National Embarrassment," *Forbes*, May 1, 2015, www.forbes.com/sites /luketimmerman/2015/05/01/under-reporting-of-severe-drug-side-effects-a -national-embarrassment/?sh=434a222b251d.

18. David Kessler, "Introducing MedWatch: A New Approach to Reporting Medication and Device Adverse Effects and product Problems," *Journal of the American Medical Association*, Vol. 269, No. 21, June 23, 1993, www .fda.gov/downloads/Safety/MedWatch/UCM201419.pdf.

19. *Bruesewitz v. Wyeth Laboratories*, United States Supreme Court, 562 U.S. (February 22, 2010), Sotomayor dissent, p. 1.

Chapter Five

1. "Scientific Review of Vaccine Safety Datalink Information," June 7–8, 2000, Simpsonwood Retreat Center, Norcross, Georgia, Centers for Disease Control and Prevention, (Accessed February 7, 2016) p. 1–2, www .thinktwice.com/simpsonwood.pdf.

2. Ibid. at 2.

3. Ibid. at 3.

4. Ibid.

5. Ibid.

6. Ibid. at 7.

7. Ibid. at 26.

8. Ibid. at 28.

9. Ibid.

10. Ibid. at 32.

11. Ibid at 36.

12. Ibid. at 42

13. Ibid. at 46.

Chapter Six

1. "Pentagon Papers," *Encyclopedia Britannica*, February 19, 2020, (Accessed August 17, 2021), www.britannica.com/topic/Pentagon-Papers.

2. Ibid.

3. Email from Dr. Brian Hooker to Dr. William Thompson, November 9, 2013.

4. Brian Hooker Notes on Conversation with Dr. William Thompson, November 13, 2013.

5. Kent Heckenlively, *Inoculated: How Science Lost Its Soul in Autism*, (New York, NY: Skyhorse Publishing 2021), 11.

6. Ibid. at p. 14–15

7. Hooker and Wakefield Complaint to CDC and Office of Research Integrity at Health and Human Services, October 14, 2014, p. 16.

8. "Statement of William M. Thompson, PhD, Regarding the 2004 Article Examining the Possibility of a Relationship Between MMR Vaccine and Autism," August 27, 2014, www.mroganverkamp.com/statement—of william-w-thompson-ph-d-regarding-the-2004-article-examining-the -possibility-of-a-relationship-between-mmrvaccine-and-autism/.

9. "Congressman William Posey on CDC Whistleblower," C-Span, July 29, 2015, www.c-span.org/video/?c455834/user-clip-bill-posey-cdc -whistleblower.

10. Sharyl Attkisson, "How a Pro-Vaccine Doctor Reopened Debate About Link to Autism," *The Hill*, January 13, 2019, www.hill.com/opinion /healthcare/425061-how-a-pro-vaccine-doctor-reopened-debate-about-link -to-autism.

11. Ibid.

Chapter Seven

1. Amazon Charts for Non-Fiction, May 10, 2020, www.amazon.com/charts /2020-05-10/mostsold/nonfiction?ref=chrt_bk_nav_back

2. Michelle Malkin Twitter feed, April 16, 2020.

3. Garson and Wright Public Relations Services and Literary Management Services, www.gwprbooks.com/wright-literary-book-management—llc. html (Accessed November 29, 2020.)

4. Aimee Picchi, "Drug Ads: $5.2 Billion Annually—And Rising," *CBS News*, March 11, 2016, www.cbsnews.com/news/drug-ads-5-2-billion -annually-and-rising/

5. Ibid.

6. Ibid.

7. Reenita Das, "Are Direct to Consumer Ads for Drugs Doing More Harm than Good?", Forbes, May 14, 2019, www.forbes.com/sites/reenitadas /2019/05/14/direct-to-consumer-drug-ads-are-they-doing-more-harm-than -good/?sh=e3397324dfc6

8. Joanne Kaufman, "Think You're Seeing More Drug Ads on TV? You Are, and Here's Why," the *New York Times*, December 24, 2017, www.nytimes .com/2017/12/24/business/media/prescription-drugs-advertising-tv.html

9. John Vibes, "Robert F. Kennedy, Jr, Says 70% of News Advertising revenue Comes from Big Pharma," *True Activist*, June 1, 2015, www .trueactivist.com/robert-f-kennedy-jr-says-70-of-news-advertising -revenue-comes-from-big-pharma-t1/

10. Karl Evers-Hillstrom, "Big Pharma Continues to Top Lobbying Spending," *Open Secrets*, October 25, 2019, www.opensecrets.org/news /2019/10/big-pharma-continues-to-top-lobbying-spending/

11. Ibid.

12. Ibid.

13. "Total Global Vaccine Market Size from 2014–2024," Statista, (Accessed November 30, 2020), www.statista.com/statistics/943867/size-of-the -global-vaccine-market/

14. Davey Alba, "Virus Conspiracists Elevate a New Champion," the *New York Times*, May 9, 2020, www.nytimes.com/2020/05/09/technology /plandemic-judy-mikovitz-coronavirus-disinformation.html.

15. Ibid.

16. Ibid.

17. Ibid.

18. Ibid.

19. Ibid.

20. Ibid.

21. EJ Dickson, "Judy Mikovits, Disgraced Doctor at the Center of 'Plandemic,' Has a Bestselling Book on Amazon," *Rolling Stone*, May 12, 2020, www .rollingstone.com/culture/culture-news/plandemic-judy-mikovits-plague -of-corruption-998224/

22. Ibid.

23. Ibid.

24. Ibid.

25. Ibid. Comment by Goldweights, May 20, 2020

26. Ibid. Comment by Orange Coathanger, May 23, 2020

27. Jennifer Kasten, "What Judy Mikovits Gets Wrong: Pretty Much Everything, Pathologist Argues," *MedPage Today*, May 12, 2020, www .medpagetoday.com/infectiousdisease/generalinfectiousdisease/86461

28. Mikhail Varshavski, "Point by Point 'Plandemic' Smackdown," *MedPage Today*, May 14, 2020, www.medpagetoday.com/infectiousdisease /covid19/86487

29. Paula Froelich, "Who is Judy Mikovits, the Woman Behind Banned 'Plandemic' Conspiracy Video?", *New York Post*, May 16, 2020, www .nypost.com/2020/05/16/who-is-judy-mikovits-the-woman-behind-the -banned-plandemic/

30. Ibid.

31. Ali Breland, "After a Conspiracy Site Boosted the Debunked 'Plandemic' Video, Trump Pushed Its Take on Joe Scarborough," *Mother Jones*, May 28, 2020, www.motherjones.com/politics/2020/05 /trump-joe-scarborough/.

32. "Meet the Press Transcripts," *Meet the Press*, May 17, 2020, www.nbcnews
    .com/meet-the-press/meet-press-may-17-2020-n1208626.

33. Ibid.

34. Ibid.

35. Ibid.

36. Ibid.

37. Ibid.

38. "Malcom X Quotes," Good Reads, (Accessed September 1, 2021), www
    .goodreads.com/quotes/280633-if-you-re-not-careful-the-newspapers
    -will-have-you-hating.

Chapter Eight

1. Kimiko de Freytas-Tamura. "A Hospital Finds an Unlikely Group
   Opposing Vaccination: Its Workers," *New York Times*, August 22, 2021,
   www.nytimes.com/2021/08/22/nyregion/staten-island-covid-vaccine
   -workers.html.

2. Ibid.

3. Ibid.

4. Ibid.

5. Lena Howland, "'We Just Don't Want to be Forced' Why Some
   Healthcare Workers are Pushing Back Against Vaccine Mandate," *ABC
   10*, August 16, 2021, www.abc10.com/article/news/health/coronavirus
   /some-healthcare-workers-push-back-vaccine-mandate/103-a91df4f7-e6e8-
   45e7-9f2b-06209b1e8a93.

6. Brandon Tydry, "People in Nebraska City Gather for Protest Against
   Vaccine Mandates," *6 News on Your Side*, August 21, 2021, www.wowt
   .com/2021/08/21/people-nebraska-city-gather-protest-against-vaccine
   -mandates/.

7. Ibid.

8. Danielle Wallace, "Chicago Police Union '100% Against' Vaccine
   Mandates for Officers: 'Hell, no,'" Fox News, August 25, 2021, www
   .foxnews.com/us/chicago-police-against-vaccine-mandate.

9. "For Fourth Week, French Protest Health Pass, Vaccine Mandate," *VOA
   News*, August 7, 2021, www.voanews.com/covid-19-pandemic
   /fourth-week-french-protest-health-pass-vaccine-mandate.

10. Shannon Thaler, "Americans with PhDs are the Most Reluctant to Get
    Vaccinated Against COVID, Study Finds," *Daily Mail*, August 14, 2021,
    www.dailymail.co.uk/news/article-9893465/Americans-PhDs-reluctant
    -vaccinated-against-COVID-study-finds.html.

11. "Phase 3 Trial Shows Regen-Cov (Casirivimab with Imdevimab) Antibody
    Cocktail Reduced Hospitalization or Death by 70% in Non-Hospitalized

Covid 19 Patients," *Regeneron Press Release*, March 23, 2021, www
.newsroom.regeneron.com/news-releases/news-release-details/phase
-3-trial-shows-regen-covtm-casirivimab-imdevimab-antibody.

Epilogue

1. Jocelyn Gecker and Kathleen Ronayne, "California Requires Vaccines,
    Tests for Teachers and Staff," *Associated Press*, August 11, 2021, www
    .apnews.com/article/health-education-california-coronavirus-pandemic
    -7699f4496101d261ac94ed22e7544417.

Acknowledgments

I'd first like to thank my wonderful partner in life, Linda, and our two children Jacqueline and Ben, for their constant love and support. I'd like to thank my mother, Josephine, and my father, Jack, for teaching me to tell the truth regardless of the consequences and showing how to love through even difficult times. I'd like to thank the best brother in the world, Jay, and his wonderful wife, Andrea, and their three children, Anna, John, and Laura, for always being on my side.

I'd like to thank some of the wonderful teachers in my life, my seventh- grade science teacher, Paul Rago, my eighth-grade English teacher, Elizabeth White, my high school science teacher, Ed Balsdon, my religion teacher Brother Richard Orona, and in college, English professors Clinton Bond, Robert Haas, Carol Lashoff, and in the political science department, David Alvarez, who nominated me to be the school's Rhodes Scholar candidate. I'd also like to thank my college rowing coach, Giancarlo Trevisan, the mad Italian, who showed me what it means to have crazy passion for an often-overlooked sport. In law school, I'd like to thank Bernie Segal, the criminal defense attorney who taught me to always have hope that justice will eventually prevail. I'd like to thank my writing teachers, James Frey, who looked at me one time and said, "Yeah, I think you'll be a writer," as well as Donna Levin, and James Dalessandro, who always said to find the story first, then write the hell out of it.

My life wouldn't be complete without my great friends, John Wible, John Henry, Pete Klenow, Chris Sweeney, Suzanne Golibart, Gina

Cioffi, Eric Holm, Susanne Brown, Rick Friedling, Max Swafford, Sherilyn Todd, Rick and Robin Kreutzer, Christie and Joaquim Pereira, Tricia Mangiapane, and all of you who have made my passage through life such a party.

I work with the best group of science teachers at Gale Ranch Middle School: Danielle Pisa, Neelam Bhokani, Amelia Larson, Matt Lundberg, Katie Strube, Derek Augarten, and Arash Pakhdal. Thanks for always challenging my thinking and making me ask what is best for our students.

In the activist community, I'd like to thank J. B. Handley, Tim Bolen, Mary Holland, Lou Conte, Del Bigtree, Brian Hooker, Barry Segal, Elizabeth Horn, Brian Burrowes, Polly Tommey, Dr. Andrew Wakefield, and Robert F. Kennedy Jr., for their friendship in the continuing fight against the Goliath of corrupted science.

Lastly, I'd like to thank my agent, Johanna Maaghul, my wonderful editors, Anna Wostenberg; and at Skyhorse, the fabulous Caroline Russomanno, and for the faith shown in me over the years by publisher Tony Lyons.

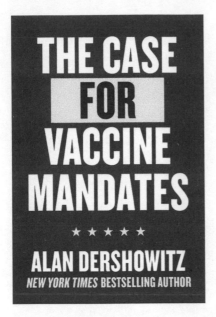

For the opposite argument, please read *The Case for Vaccine Mandates* by Alan Dershowitz.

In *The Case for Vaccine Mandates*, Alan Dershowitz—*New York Times* bestselling author and one of America's most respected legal scholars—makes an argument, against the backdrop of ideologically driven and politicized objections, for mandating (with medical exceptions) vaccinations as a last resort, if proved necessary to prevent the spread of COVID.

Alan Dershowitz has been called "one of the most prominent and consistent defenders of civil liberties in America" by *Politico* and "the nation's most peripatetic civil liberties lawyer and one of its most distinguished defenders of individual rights" by *Newsweek*. He is also a fair-minded and even-handed expert on civil liberties and constitutional rights, and in this book offers his knowledge and insight to help readers understand how mandated vaccination and compulsion to wearing masks should and would be upheld in the courts.

The Case for Vaccine Mandates offers a straightforward analytical perspective: If a vaccine significantly reduces the threat of spreading a serious and potentially deadly disease without significant risks to those taking the vaccine, the case for governmental compulsion grows stronger. If a vaccine only reduces the risk and seriousness of COVID to the vaccinated person but does little to prevent the spread or seriousness to others, the case is weaker. Dershowitz addresses these and the issue of masking through a libertarian approach derived from John Stuart Mill, the English philosopher and political economist whose doctrine he summarizes as, "your right to swing your fist ends at the tip of my nose." Dershowitz further explores the subject of mandates by looking to what he describes as the only Supreme Court decision that is directly on point to this issue; decided in 1905, *Jacobson v. Massachusetts* involved a Cambridge ordinance mandating vaccination against smallpox and a fine for anyone who refused.

In the end, *The Case for Vaccine Mandates* represents an icon in American law and due process reckoning with what unfortunately has become a reflection of our dangerously divisive age, where even a pandemic and the responses to it, divide us along partisan and ideological lines. It is essential reading for anyone interested in a non-partisan, civil liberties, and constitutional analysis.